IMAGES OF WAR

SPECIAL FORCES VEHICLES

VEHICLES

1940 TO THE PRESENT DAY

IMAGES OF WAR

SPECIAL FORCES VEHICLES

1940 TO THE PRESENT DAY

RARE PHOTOGRAPHS FROM WARTIME ARCHIVES

Pat Ware

Pen & Sword
MILITARY

First published in Great Britain in 2012 by
PEN & SWORD MILITARY
an imprint of
Pen & Sword Books Ltd,
47 Church Street,
Barnsley,
South Yorkshire
S70 2AS

A CIP record for this book is available from the British Library.

ISBN 978 1 84884 642 5

Typeset by Chic Media Ltd.

Printed and bound by CPI Group (UK) Ltd, Croydon, CR0 4YY

Pen & Sword Books Ltd incorporates the Imprints of
Pen & Sword Aviation, Pen & Sword Family History, Pen & Sword Maritime, Pen & Sword Military, Pen & Sword Discovery, Wharncliffe Local History, Wharncliffe True Crime, Wharncliffe Transport, Pen & Sword Select, Pen & Sword Military Classics, Leo Cooper, The Praetorian Press, Remember When, Seaforth Publishing and Frontline Publishing.

For a complete list of Pen & Sword titles please contact
Pen & Sword Books Limited
47 Church Street, Barnsley, South Yorkshire, S70 2AS, England
E-mail: enquiries@pen-and-sword.co.uk
Website: www.pen-and-sword.co.uk

Contents

Introduction

Anyone who has read of the exploits of the SAS in the Western Desert during 1941/42 or who perhaps watched the black-hooded Special Projects Team, led by Captain Dick Arthur, storm the Iranian Embassy back in 1980 will have nothing but admiration for the capabilities of today's highly sophisticated Special Forces. And the British SAS doesn't have a monopoly on this type of mission. Witness the rescue of hostages from Mogadishu by German Special Forces in 1977, or the Israeli paratroopers who similarly rescued hostages during the famed raid on Entebbe the previous year.

The tactics and role of the modern Special Forces unit can be traced back to the earliest days of the Second World War when German, Italian and British commando units carried out missions, frequently behind enemy lines, which invariably carried an unprecedented degree of risk and danger. A key element of these operations was often the self-contained nature of the forces involved, who were able to operate outside the normal military command structure.

Following the end of that conflict the Special Forces, far from being disbanded, became increasingly important as the more powerful nations began to engage in what has subsequently become known as asymmetric warfare. It was found that small, well-trained units, often operating covertly, were frequently able to achieve objectives that were not possible using conventional forces. Similarly Special Forces proved very useful in operations where the element of secrecy was important and where it might subsequently be necessary to invoke a degree of political deniability. Such forces have played a role in many of the post-war conflicts, including the wars in Indo-China, the troubles in Northern Ireland, the Falkland Islands, the Balkans, Chechnya, the former Yugoslavia and in the Middle East. During the US-led invasion of Afghanistan Special Forces from several countries played a major part in dislodging the Taliban from power.

The typical Special Forces soldier is well trained, highly motivated and well equipped, and will almost certainly have spent some time in a regular unit before putting himself forward for what is, inevitably, an arduous selection and induction course. Working in small groups, Special Forces are trained to expect the unexpected and to deal with it, often making decisions on the spot without recourse to a complicated command structure. Many Special Forces operations involve deep penetration into enemy-held territory, or lengthy reconnaissance missions, either gathering intelligence or destroying enemy assets. Crucial to the success of this type of operation is the selection of the right type of vehicle. It hardly needs to be said that reliability and durability are a given, but the chosen vehicle must also be able to carry whatever equipment is deemed necessary for the mission, including weapons, ammunition, fuel, food, water and personal kit, as well as a selection of spare parts and replacements for possible repair in the field.

The Long Range Desert Group (LRDG), which might be considered to be the first modern Special Forces unit, chose to operate modified Chevrolet civilian trucks, but it was the stripped-down Jeeps of the SAS that established the norm for this type of operation. When these Jeeps wore out they were replaced by Series I Land Rovers, which in turn were superseded by the iconic Series IIA Land Rover 'Pink Panthers'. These were the first vehicles to be constructed by an outside contractor – in this case Marshalls of Cambridge – to the requirements of the SAS Regiment and they remained the pattern for Special Forces' vehicles until the appearance of dune buggy-based fast-strike vehicles in the early 1980s; in Afghanistan, these have subsequently been replaced by larger, armoured vehicles such as the Jackal.

Although all the vehicles shown in this book are authentic, the photographs include both re-enactors and static shots taken, for example, at defence exhibitions. The reason for this should be obvious, since it is in the nature of clandestine operations that pictures are rarely taken in action.

The stripped-down, overloaded Jeeps so typical of the SAS Regiment and other Special Forces units operating in the desert during the Second World War should probably be considered as the grandfather of all such vehicles. Here, a 'Greek Sacred Company', manned by officers who had escaped from Greece and Crete, sets out on desert patrol in a column of what were described at the time as 'special commando Jeeps'. (IWM, E23148)

Based on the Land Rover Series IIA, the FV18064 was a heavily modified Special Forces patrol vehicle developed by Marshalls of Cambridge and the SAS Regiment. It was designed to withstand the harshest operating conditions and carried whatever stores and equipment were required to allow the crews to operate for long periods away from a base; the nickname 'Pink Panther' was derived from the distinctive paint finish adopted for vehicles operating in Oman. (*Simon Thomson*)

The Supacat 'high mobility transporter' (HMT 400) or Jackal – seen here in upgraded Jackal 2 configuration – has recently been adopted by the SAS as a replacement for the long-wheelbase Land Rovers favoured by the regiment for the last fifty years and is already seeing service in Afghanistan. (*Sergeant Mike Fletcher; MoD, Crown copyright*)

Chapter One

The Role of Special Forces

Although the concept of 'Special Forces' – meaning military units that have been highly trained and specially equipped in order to undertake unconventional operations – isn't new, the recent rise in so-called asymmetric warfare has seen an increase in the use of Special Forces for operations such as body-guarding, hostage release, infiltration of terrorist groups, reconnaissance and intelligence gathering, demolition, and interdiction and harassment of enemy resources. Most Special Forces units are manned by volunteers, elite personnel who are physically and mentally robust, and who have been put through an arduous selection and training regime designed to eliminate all but the toughest and fittest men. During training, emphasis is placed on the ability of the men both to work as a team and to operate individually if necessary; great importance is placed on the need for resourcefulness. Usually considered to be 'high-value assets', Special Forces tend to be commanded at a strategic level and are generally able to deliver results that are disproportionate to their size.

As far back as 3,000 years ago the Chinese strategist Jiang Ziya advocated the deployment of well-trained elite troops to gain advantage over an enemy, and the Romans used hand-picked men for scouting and commando-style missions. More recently the British Army established specialised 'scouting units' during the second Anglo-Boer War (1899–1902), recruiting skilled woodsmen who were well practised in fieldcraft and marksmanship.

Modern Special Forces started to emerge in the early twentieth century – Lawrence of Arabia's operations would certainly fall within any modern definition of Special Forces' actions – but it wasn't until the Second World War that the real value of such units came to be recognised in conducting clandestine operations behind enemy lines, gathering intelligence, disrupting fuel and ammunition supply logistics, supporting and training insurgents and resistance groups, and engaging in various offensive actions. The wartime exploits of groups such as the British SAS, Popski's Private Army and the Long Range Desert Group have long since become the stuff of legend, and the SAS went on to become the model upon which other similar military units were based. Before the war was over there were SAS squadrons operating under French, Greek and Belgian flags, and within another decade similar units had also been formed in the USA and in New Zealand, Australia and Israel.

Special forces have become increasingly important in the last few decades. As well as playing a significant role in combating terrorism around the world, such units have also made a significant contribution to conventional military operations. In the recent conflicts in Kosovo, Iraq and Afghanistan, for example, NATO Special Forces were often deployed to work alongside local guerrilla or rebel forces, providing training and support, and exploiting their in-depth knowledge of local terrain and enemy strengths and movements. During the summer of 2011 US Special Forces were credited with the operation against Osama Bin Laden in Pakistan.

For the modern Special Forces soldier the selection procedures are rigorous. Only the very best applicants are selected and even then there tends to be a very high rate of drop-out during the initial training period. Huge emphasis is placed on peak physical fitness and teamwork, and men are encouraged to develop special skills appropriate to their particular unit. The training regime remains constant and intensive, and members of a team are chosen for their ability to work seamlessly together, with each team member trained to be able to carry out more than one role. The unpredictability of the typical Special Forces operation not only demands these rigorous training and operational skills, but also calls for a unique command structure. Operational orders tend to come from the very highest levels of government, while the operatives on the ground are allowed an enormous degree of autonomy as the operation unfolds, with the men often being called upon to make, and carry through, decisions that would normally be above their 'pay grade'.

The equipment selected for an operation is equally important, including the use of appropriate clothing. Weapons are also often chosen specifically for each operation, with criteria such as rate of fire, physical size and ammunition type all carefully considered. Maximum use is made of all appropriate modern technologies, with computer and surveillance equipment selected for its suitability to the task in hand.

Today's Special Forces soldiers are well trained and motivated, physically fit, mentally prepared and well equipped. With the courage and skill to operate in small teams or as individuals, often in isolation or well inside enemy-held territory, these men can truly claim to be among the world's elite soldiers.

It was the British Long Range Desert Group (LRDG) that developed the tactic of using heavily modified vehicles to carry out extended operations in the Western Desert, at first using modified civilian Chevrolets but subsequently using columns of Jeeps. (*Simon Thomson*)

Although best-known for their distinctive sand-coloured Jeeps, which were devised for desert operations, the SAS also operated in northwest Europe following the invasion of Normandy in 1944. Note the additional fuel tank in the rear, protective windscreens and drum-fed Vickers K machine guns. (*Tank Museum*)

Opposite, top: The American high-mobility multi-purpose wheeled vehicle (HMMWV) can be easily adapted to suit many roles and is operated by all branches of the US military. Here US Marines are about to begin a live-firing exercise in Morocco during Exercise African Lion 2007. (*Corporal Dustin T. Schalue; US DoD*)

Opposite, bottom: Manufactured by Babcock International, the six-wheeled Supacat Coyote TSV is operated alongside the smaller Jackal as a supply vehicle. The photograph shows the vehicle during training. (*Andrew Linnett; MoD, Crown copyright*)

During the 1980s and 1990s the traditional type of Special Forces vehicle started to be supplanted by what were essentially militarised dune buggies. Here we see a pair of heavily armed Chenowth fast attack vehicles (FAV) operated by US Navy SEALs and the US Marines. The prominent dome conceals complex communications equipment that allows the crews to communicate with one other while riding in the vehicles. (*Photographer's Mate 1st Class (SW) Arlo Abrahamson; US Navy*)

Opposite, top: A real favourite of modern re-enactors, the SAS Jeep was stripped of unnecessary weight and modified to allow reliable operation in harsh desert terrain. Note the cut-back radiator grille and the condensing system designed to save precious water. The weapons are Vickers K observer's machine guns on a twin mount. (*Simon Thomson*)

Opposite, bottom: Modelled on SAS Jeeps and the subsequent Series I Land Rovers (FV18000 series) that were adapted in much the same way, the Minerva Special Forces vehicle was a licence-built Land Rover designed for Belgian parachute commandoes. On-board weapons generally included three FN MAG 7.62mm machine guns. (*Simon Thomson*)

Ever since adopting the Series I Land Rover for the Special Forces role back in 1955, the British Army has tended to favour the products of Solihull. With its high-capacity pick-up bed, the long-wheelbase Defender desert patrol vehicle (DPV) seen here (above and below) was developed by Marshalls of Cambridge and started to enter service in the 1980s. Sand channels are a must on any vehicle likely to venture into desert territory. (*Roland Groom; Tank Museum*)

Not all Special Forces operations are conducted using the vehicles generally associated with this role. Developed in the mid-1970s as an anti-tank vehicle, this Panhard ERC-90 (*engin de reconnaissance à canon de 90mm*) SAGAIE is in service with a reconnaissance squadron of the Foreign Legion, and was photographed near Djibouti in 2005. (*Davric*)

Developed by the Longline Company around a high-tensile tubular-steel space-frame, the Cobra light strike vehicle (LSV) was designed as a highly mobile weapons platform for use in hostile environments. A fast and reliable vehicle, it was adopted by the SAS during the Gulf War. (*Simon Thomson*)

Despite its small size and minimal creature comforts, the six-wheeled Supacat all-terrain mobile platform (ATMP) is a versatile machine that is well-suited to the Special Forces role. The example seen here has just fired a MILAN anti-tank missile. (*Warehouse Collection*)

Opposite, top: Like the SAS Jeep and the 'Pink Panther', the SAS desert patrol vehicle (DPV) is popular with military vehicle collectors; with all of the original kit, and with the faces of the crew suitably concealed, this could almost be the real thing. (*Simon Thomson*)

Opposite, bottom: A US Marine of Echo Company, 2nd Battalion, 8th Marine Regiment approaches an up-armoured HMMWV in Ramadi, Iraq, in 2008. The armoured variant of the HMMWV was developed by O'Gara-Hess & Eisenhardt following the attempt to rescue downed helicopter crews in Somalia in 1993. (*Corporal Jeremy M. Giacomino; US DoD*)

Heavily laden Land Rover Wolf XD Defender patrol vehicles of the British Army crossing a temporary bridge in Iraq. The 'weapons mount installation kit' (WMIK) typically fitted to the Special Forces Wolf was developed in early 1998 by Ricardo Special Vehicles; the installation also includes a package of other automotive improvements. (*Warehouse Collection*)

The US Marines light armoured vehicle (LAV-25) is not typical of vehicles used in the Special Forces role. This example, photographed in Helmand Province, Afghanistan, is with US Marines of Delta Company, 2nd Light Armored Reconnaissance Battalion and is providing security support to local troops. (*Corporal Charles T. Mabry II; US DoD*)

Bristling with weapons and stuffed to the gunwales with personal kit, food and other supplies, the iconic 'Pink Panther' typifies Special Forces vehicles. In fact, just seventy-two examples were constructed and the vehicle has subsequently become highly collectable, with enthusiasts hunting down all the correct on-board equipment. (*Simon Thomson*)

US Marines from the Security Cooperation Task Force (SCTF) disembarking an HMMWV from a landing craft in Covenas, Sucre, Colombia, in January 2011. The SCTF was on a scheduled deployment to support Amphibious Southern Partnership Station 2011, combining naval and amphibious operations to assist the host nation in stability and peacekeeping operations. (*Mass Communication Specialist 3rd Class Lauren G. Randall; US Navy*)

Equipped with the WMIK combined roll-cage and weapons mount, this British Army Land Rover Wolf XD Defender is armed with a 0.50in heavy machine gun in the rear and has a 7.62mm general-purpose machine gun (GPMG) on a pintle mount in front of the co-driver's seat. (*Warehouse Collection*)

During 1942 the Long Range Desert Group started operations using a total of 200 modified Canadian Chevrolet 1533X2 patrol vehicles alongside a ragbag of other modified Ford and Chevrolet trucks. The LRDG remains a popular theme with British re-enactor groups. (*Simon Thomson*)

A WMIK-equipped Land Rover Wolf XD Defender seen at the Defence Vehicle Dynamics show in 2006. These vehicles were widely used for patrols in Iraq and Afghanistan but offered little protection for the crew, and a spate of fatal incidents has led to the deployment of more heavily armoured vehicles. (*Warehouse Collection*)

Developed to replace the ageing Land Rover WMIK, the Supacat Jackal is not typical of previous Special Forces machines, but offers increased carrying capacity without sacrificing speed or mobility; add-on armour can be fitted to enhance crew survivability. (*Simon Thomson*)

Photographed at Camp Viper in Basra, this Land Rover Wolf XD Defender of 16 Air Assault Brigade has been fitted with the WMIK combined roll-bar and weapons mount, and is typical of vehicles operating on patrols in Iraq. (*Master Sergeant Edward D. Kniery; US DoD*)

A privately owned ex-US Marine Corps Chenowth fast attack vehicle (FAV), mounting a pair of heavy machine guns and sporting the distinctive communications dome. Fast and light, the FAV offered considerable advantages over the slower HMMWV. (*Simon Thomson*)

Resplendent in its matt pink paintwork, this privately owned 'Pink Panther' is said to be in 'original' ex-services condition. (*Warehouse Collection*)

Chapter Two

Special Forces Units

M any of today's elite special units are modelled on the training and tactics of the SAS, but, alongside intelligence gathering and typical 'hit and run' raids, the Special Forces' role has also expanded to include counter-terrorism and the training and coordination of local troops. At the same time the use of advanced technology – so-called 'effects-based warfare' – has broadened the role and scope of Special Forces operations, allowing them to act as pathfinders for larger-scale operations and to operate alongside conventional troops when necessary.

The armed forces, or security police services, of most developed nations include Special Forces units, and a list of such units would include more than seventy names, with almost half of them attached to the British and US military. Other nations with designated Special Forces units include Argentina, Australia, Austria, Canada, China, Colombia, Denmark, El Salvador, France, Germany, India, Israel, Italy, Jordan, the Netherlands, New Zealand, Norway, Poland, Russia, South Africa, South Korea and Spain. Typical roles include airborne assault, counter-terrorism, hostage rescue, VIP protection, foreign unit training, reconnaissance and intelligence gathering . . . and even state-sponsored assassination. Many are also tasked with fighting organised crime and drug-running operations. Some of these units, including a number that are no longer in existence, are described here, but details such as current strengths are not necessarily in the public domain.

AUSTRALIAN SAS
Motto: Who dares wins
Formed in 1957 during the insurrection in Malaya, the Australian SAS – originally designated the 1st Special Air Service Company but renamed the Special Air Service Regiment (SASR) in 1964 – has also seen service in Afghanistan, Borneo, Brunei, Iraq, Kuwait and Vietnam. It currently comprises six squadrons, three of which are combat units; one of the latter is always on standby for counter-terrorism operations.

The SAS is open to volunteers from any part of the Australian Defence Force, but the current strength of the regiment remains classified.

BELGIAN SAS
Motto: Who dares wins

Originally formed at Malvern Wells in Worcestershire in January 1942 as the 1st Belgian Independent Parachute Company, it was subsequently renamed the 1st Belgian SAS Squadron and became part of the British SAS Brigade in 1944, where it was referred to as 5th SAS. The unit's role was primarily one of sabotage, reconnaissance and intelligence gathering, and it saw its first action in July 1942 when members were parachuted behind enemy lines into occupied France. In August 1944 several teams were dropped near the Belgian border in order to make contact with resistance units in Belgium; during the Ardennes offensive in the winter of 1944/45 the unit was equipped with armoured Jeeps.

On 21 September 1945 5th SAS was transferred to the newly re-formed Belgian Army and was renamed the 1st Regiment of Parachutists. In 1952 the unit ceased to be independent, becoming 1 PARA Battalion, Paracommando Regiment; it was planned that 1 PARA would be disbanded during 2011.

Post-war operations included counter-insurgency during the Simba uprising in the Belgian Congo, as well as deployment in Burundi, Rwanda and Somalia.

FRENCH SPECIAL FORCES
Foreign Legion
Motto: *Legio patria nostra* – The Legion is our Fatherland or *Marche ou crève* – March or die (unofficial)

The *Légion Etrangère* – or French Foreign Legion – was established in 1831 by the French King Louis-Philippe I and at first was manned exclusively by foreign nationals willing to serve in the French armed forces. Today it is equally open to French citizens.

The Legion was originally stationed in Algeria, which it helped to pacify and develop, and was traditionally intended to fight outside France, being used to expand and defend the French Empire during the nineteenth century, as well as fighting in the Franco-Prussian wars and in both world wars. Following the end of the Second World War large numbers of ex-*Wehrmacht* soldiers and many other veterans joined the Legion, recognising that this was an opportunity to pursue an elite military career of the sort that was no longer available in their homeland; Germans continue to constitute a strong presence in the Legion, which today numbers around 7,500 men.

In the post-war years the Legion has become a unique elite military unit with a very strong esprit de corps and a formidable reputation, and has fought in Indo-China, in the Algerian War, in Chad, Zaire and the Gulf War. In 1962, following the independence of Algeria, the Legion was downsized and relocated to France, with its headquarters at Aubagne.

1ère Régiment de Parachutistes d'Infanterie de Marine

Motto: *Qui ose gagne*; Who dares wins

Originally formed in 1941 in Britain, where it was designated the 3rd and 4th (French) SAS, the French *1ère Régiment de Parachutistes d'Infanterie de Marine* (RPIMa) was composed mainly of French citizens. The units were active in metropolitan France, notably in Brittany and Normandy.

Following the end of the Second World War 3rd and 4th SAS were handed over to the French Army, becoming known as the *1ère Compagnie d'Infanterie de l'Air*, before being redesignated as the *1ère Compagnie de Chasseurs Parachutistes*, and then the *1ère Régiment de Parachutistes d'Infanterie de Marine* (RPIMa). Today, the RPIMa consists of four companies, with headquarters in Bayonne, and is one of the regiments of the French Army Special Forces Brigade. Tactics and training are based heavily on British SAS practice, and the unit has seen action in Indo-China and Libya.

Groupe d'Intervention Gendarmerie Nationale

Formed in 1974, the *Groupe d'Intervention Gendarmerie Nationale* (National Gendarmerie Intervention Group, or GIGN) is a small, highly trained, anti-terrorist police unit specialising in hostage release and the suppression of serious civil unrest and disturbance. Great emphasis is placed on marksmanship, and the unit always includes a number of crack snipers.

Since its formation, the unit has taken part in more than 700 actions, and rescued around 500 hostages; notable successes include the rescue of hostages at Marseilles airport in 1994. The unit cross-trains with other similar forces, including the FBI and Delta Force, and assisted in training the Saudi National Guard for their assault on the Grand Mosque in 1979.

GERMAN SPECIAL FORCES

During the Second World War highly trained German commando Special Forces, known as *Brandenburgers*, were established by the *Abwehr* (military intelligence) in October 1939 under the command of Admiral Canaris. The Brandenburgers assisted the German Army during the *Blitzkrieg* campaigns in Belgium, France and the Netherlands, and seized the docks at Orsova on the River Danube in spring 1941. A second Special Forces unit, known as Sonderverbände, was formed in April/May 1941 and trained in Greece before seeing operations in Tunisia, the Caucasus and the Balkans. Some 10,000 of General Kurt Student's paratroopers were dropped into Crete where they took and held the island during eight days of fierce fighting. In September 1943 German paratroops freed Mussolini when he was being held by Italian partisans, and paratroop operations continued until the end of the war.

In more recent times the German GSG-9 (*Grenzschutzgruppe 9*) unit was created following the massacre of eleven Israeli athletes at the 1972 Olympics in Munich. With three combat units totalling 250 men, GSG-9 was originally controlled by the state border police; today it is under the control of the Federal Police, and remains Germany's primary counter-terrorist unit. The unit has been involved in many operations, but the best known of these is the rescue of hostages being held by the Popular Front for the Liberation of Palestine at Mogadishu in 1977; all the hostages and the crew of the airliner were rescued alive.

BRITISH SPECIAL FORCES
16 Air Assault Brigade
Motto: Get the job done!
Formed in 1999 following a Strategic Defence Review, 16 Air Assault Brigade is a rapid response airborne unit, and was created by amalgamating elements of 5 Airborne Brigade and 24 Airmobile Brigade. While perhaps not universally regarded as 'Special Forces', 16 Air Assault Brigade, with 8,000 personnel, is the British Army's premier rapid response formation, and is able to bring together the agility and reach of airborne forces with the potency of the attack helicopter. Capabilities include air assault infantry, parachute, signals, pathfinders, attack and utility aviation, artillery, engineers, armoured reconnaissance, logistics and equipment support, and medical and provost capabilities.

The Brigade is a light, adaptable and potent force, able to pack a heavy punch wherever required, and has already seen action in Afghanistan, Iraq, Macedonia and Sierra Leone.

Chindits
Motto: The boldest measures are the safest!
Formed and led by Major General Orde Wingate DSO, the 77th Indian Infantry Brigade and latterly the 3rd Indian Infantry Division – both units being better known as the Chindits – were the largest of the Allied Special Forces of the Second World War. The Chindits, whose name was derived from Chinthe, a mythical beast and guardian of Burmese temples, were long-range penetration units operating on foot, with each man carrying more than 72 pounds of equipment, and relying on surprise to target enemy lines of communication. The men were supplied by stores parachuted from transport aircraft, and close air support was used as a substitute for heavy artillery.

The first of two Chindits forays into Burma was Operation Longcloth in February 1943, when a total of 3,000 men marched over 1,000 miles in an operation designed to disable the Burmese railway system. During the campaign, which was scarcely a success, a total of 818 men were killed, taken prisoner or died of disease, which represents close to one-third of the original force. The second expedition, Operation

Thursday, carried out during February and March 1944, was the second largest airborne invasion of the war. A force of 20,000 British and Commonwealth soldiers, with air support provided by the 1st Air Commando USAAF, was assigned to assist in pushing the Ledo Road through northern Burma to link up with the Burma Road, thus creating a northern supply route to China. Wingate's strategy called for the creation of fortified bases behind the Japanese lines, allowing raiding columns to be sent out over short distances. Wingate was killed in an air crash in March 1944 and, following heavy fighting, the remains of his exhausted force were finally withdrawn during August 1944. The Chindits were disbanded in February 1945.

Long Range Desert Group
Motto: *Non vi sed arte* – **Not by strength, but by guile (unofficial)**

Formed in June 1940 by Major Ralph Bagnold and General Archibald Wavell as the Number 1 Long Range Patrol Unit, the Long Range Desert Group (LRDG) operated as part of Britain's Eighth Army and was an intelligence gathering, reconnaissance and raiding unit ranging across the Western Desert and the Mediterranean area during the Second World War. Never numbering more than 350 personnel, the LRDG included men from Britain, Rhodesia (Zimbabwe) and New Zealand, all of whom were volunteers; the LRDG was best known for the considerable damage that it was able to inflict on the operations of Field Marshal Erwin Rommel's Afrika Corps.

In May 1943 the LRDG changed its role and was moved to the eastern Mediterranean, where it was tasked with missions in the Greek islands, Italy and the Balkans. Despite a request to move to the Far East in mid-1945, the LRDG was disbanded in August of that year.

Parachute Regiment
Motto: *Utrinque paratus* – **Ready for anything**

The Parachute Regiment – universally nicknamed the 'Paras' or the 'Red Devils' – was formed during 1940 when Number 2 Commando was turned over to parachute duties. Ultimately consisting of seventeen battalions, the Paras provide the airborne infantry capability of the British Army, with one battalion permanently under the command of the Director of Special Forces in the Special Forces Support Group, and the other battalions forming part of 16 Air Assault Brigade. It is the only line infantry regiment not to have been amalgamated with another unit since 1945.

During the Second World War the regiment mounted five major parachute assault operations in North Africa, Italy, Greece, France, the Netherlands and Germany, often landing ahead of all other troops. At the end of the war the regiment was reduced to three regular army battalions based at Colchester garrison, and one Territorial. The 1st Battalion is used in support of Special Forces, the 2nd and 3rd

Battalions are deployed as air assault troops, and the 4th Battalion is held in reserve. During the post-war years the Paras have seen action in Aden, Cyprus, the Falkland Islands, the Far East, Kuwait, Malaya, Northern Ireland, Palestine and Suez. The regiment has attracted considerable adverse comment for its involvement in the 'Bloody Sunday' massacre in Londonderry in 1972.

Popski's Private Army

Officially named 'Number 1 Demolition Squadron, PPA', Popski's Private Army was formed in Cairo in October 1942 by Major (later Lieutenant-Colonel) Vladimir Peniakoff DSO, MC, a Belgian national of Russian-Jewish origin who had been decorated for his intelligence reporting and petrol-dump raiding while leading the Libyan Arab Force Commando. Peniakoff was nicknamed 'Popski' by the Intelligence Officer of the LRDG after a character in a Daily Mirror cartoon.

Popski's Private Army was the smallest of the three raiding units formed in the Western Desert during the Second World War and was also the smallest independent British Army unit, with an initial total of just twenty-three other ranks. It was run on a very informal basis, with the minimum of military formality, but undertook successful operations in Algeria, Italy, Sicily and Tunisia before being disbanded in September 1945.

Raiding Support Regiment
Motto: Quit you like men

The Raiding Support Regiment (RSR) was formed in mid-1943 to provide comparatively heavily armed support units for partisan fighters, although in its earliest days it tended to be deployed as another desert raiding force alongside the LRDG, SAS and Popski's Private Army. By October 1943 the regiment comprised five batteries, and was under the command of Lieutenant-Colonel Sir Thomas Devitt, formerly of the Seaforth Highlanders.

The RSR saw its main actions in Albania, the Dalmatian Islands, Greece and Yugoslavia, often in conjunction with guerrilla units. It was disbanded in early 1945.

Royal Marines 3 Commando
Motto: *Per mare, per terram* – By land and sea

The Royal Marines' 3 Commando Brigade is the Royal Navy's amphibious infantry unit, capable of operating independently and fighting on any terrain. Long considered to be among the finest of Special Forces, the Royal Marines have the longest basic infantry training course of any NATO combat troops and form the core component of Britain's Joint Rapid Reaction Force, on permanent readiness to deploy across the globe.

Formed in 1943 as the 3rd Special Service Brigade, using a mixture of Royal Marine and army commando units, the brigade saw service in Burma. Following the end of the Second World War the army commandos were disbanded and for a period the brigade was a strictly Royal Marines' unit; recently army personnel have once again been included to provide a mixed army and marine formation, the latter consisting of three Royal Marine battalions. The army elements include an infantry battalion, artillery regiment and engineer regiment.

Since the end of the Second World War 3 Commando has served in Afghanistan, the Falkland Islands, Iraq, Kuwait and in the Suez Canal zone.

Special Air Service (SAS) Regiment
Motto: Who dares wins!

Established in July 1941, the British Special Air Service Regiment – almost universally known, simply, as the SAS – is probably the world's best-known Special Forces unit, and with its rigorous selection and training process it has also become one of the foremost elite military units in the world. SAS operations such as the freeing of the Libyan hostages in 1980, the clandestine 'shoot and scoot' raids in Oman and Borneo in the late 1950s and early 1960s, and the successful operations in the Falkland Islands in 1982 and more recently in Afghanistan have passed into military legend, giving the SAS a deserved reputation for highly professional soldiering.

The regiment was created in July 1941 following the presentation of an unorthodox plan drawn up by David Stirling, a lieutenant in the Scots Guards. The plan, which described how small teams of highly trained men could be parachuted behind enemy lines to gain intelligence, destroy aircraft and airfields, and attack the enemy's supply and reinforcement routes, was presented to General Claude Auchinleck, the commander-in-chief of British forces in the Middle East. A phantom airborne brigade had already been established as part of a deception organisation under way in the Middle East to act as a threat to enemy planning. This deception unit was known as K Detachment, Special Air Service Brigade; in approving Stirling's strategy, Auchinleck authorised the establishment of L Detachment, SAS Brigade, consisting originally of just five officers and sixty other ranks.

The first operation undertaken by the SAS was Operation Squatter, which took place on the night of 16 November 1941. Parachuted behind enemy lines, the SAS was tasked with attacking airfields at Gazala and Timimi in support of Operation Crusader. Unfortunately, a combination of adverse weather and unexpectedly fierce enemy resistance meant that the mission was a disaster, with almost one-third of the men involved being killed or captured. For their second mission the SAS troops were transported by the Long Range Desert Group (LRDG), attacking three airfields in Libya and destroying sixty aircraft without loss.

In 1942 the unit was renamed 1st SAS Regiment; at the time it consisted of four British squadrons, one Free French squadron, one Greek squadron and the Special Boat Section (SBS). Further operations were conducted jointly with the LRDG before the SAS developed the ability to operate independently using specially stripped, modified and heavily armed Jeeps. During the remaining years of the war the SAS operated in Denmark, France, Italy, Norway and Sicily.

The regiment was disbanded in October 1946, only to be reconstituted the following year as 21 SAS Regiment, forming part of the Territorial Army and eventually consisting of three squadrons, designated A, B and C. Action followed in Borneo, Korea and Malaya, where the unit developed considerable expertise in counter-insurgency operations. In 1952 B Squadron was renamed 22 SAS, once again becoming a unit of the regular army, and in the late 1950s two squadrons were sent to Oman, where the famous 'Pink Panther' Land Rovers made their debut, and a third SAS Regiment – 23 SAS – was formed. SAS men were also involved in gathering intelligence in Northern Ireland during the 1970s, as well as becoming more involved in the struggles against terrorism. Growing expertise in undercover operations led to the creation of an SAS counter-revolutionary warfare (CRW) wing in 1975. In the last three decades the units have also seen action in Afghanistan, the Falkland Islands, Kuwait, Iraq and Sierra Leone, and more recently, in Libya.

The SAS currently comprises 22 SAS, which is a regular army unit and comprises four squadrons, and 21 and 23 SAS, both of which are Territorial Army units, and each of which comprises three squadrons.

Special Boat Service
Motto: By strength and guile
Established in 1940 and previously known as the Special Boat Squadron, the Special Boat Service (SBS) forms the third arm of the British Special Forces, and is under the joint control of the same Director as the SAS and the SRR.

Based at Poole in Dorset, the unit specialises in operations at sea and along coastlines and river networks, but has also been deployed in the Iraqi deserts and in the mountains of Afghanistan. Although formerly drawn exclusively from the Royal Marines, the SBS is now open to members of other UK regiments and services.

Special Forces Support Group
The Special Forces Support Group (SFSG) was formed on 3 April 2006 around a core of members of the 1st Battalion, Parachute Regiment. It was tasked with supporting the SAS and the SBS on operations, for example using diversionary tactics and providing protective cordons or extra firepower; the SFSG has also undertaken a training and mentoring role for foreign forces.

Headquartered at St Mathan in South Wales and consisting of four strike companies, the unit was initially composed only of personnel from the Parachute Regiment, the Royal Marines and the RAF Regiment, but it has since been opened to all British military personnel who are able to meet the selection criteria. The SFSG has been heavily involved in the present campaign in Afghanistan.

Special Reconnaissance Regiment

The Special Reconnaissance Regiment (SRR) was established on 6 April 2005 as a successor to the former 14 Intelligence Company, specifically to meet a demand for a special reconnaissance capability. It was intended to relieve the SAS and the SBS of their covert reconnaissance and surveillance roles. Personnel, both male and female, are selected from existing British units and, like the SAS, the SBS and the SFSG, the SRR comes under the command of the Director, Special Forces. Like the SAS, it is also based in Hereford.

ISRAELI SPECIAL FORCES

The elite Parachute Brigade of the Israeli Defence Forces (IDF) had its origins in Unit 101, which was created in 1953 to combat Arab terrorism but disbanded a year later and merged with other paratroop units following the outcry provoked by the Qibya massacre. Within three or four years the unit had grown to brigade size and fought with distinction in the Six-Day War against Egypt, Jordan and Syria in 1967. The brigade is currently heavily engaged in Israel's struggle against HAMAS and other radical Islamic groups.

The IDF also includes the covert *Sayeret Matkal* anti-terrorist unit under the command of the intelligence directorate, Amman. Modelled on the British SAS, the unit is involved in intelligence gathering, often conducting deep reconnaissance missions behind enemy lines. *Sayeret Matkal* is also responsible for hostage rescue operations and in 1967 carried out a successful assault on the Ugandan airport at Entebbe to rescue Jewish hostages being held by a consortium of terrorists drawn from the West German Baader-Meinhof gang and the Popular Front for the Liberation of Palestine. The unit has also been involved in several operations in Lebanon and Syria, and was even rumoured to have been behind an unsuccessful plot to assassinate Saddam Hussein in 1992.

SOVIET UNION/RUSSIAN FEDERATION SPECIAL FORCES

Motto: Any mission, any time, any place

The Soviet *Spetsnaz* units – the name is derived from *Voiska Spetsialnogo Naznacheniya,* simply meaning special-purpose troops – were formed in 1945 as spearhead troops for the Cold War, with *Spetsnaz* units established in all three of the

Soviet armed forces, as well as in the state security forces. Training, which can last up to five years, is particularly rigorous and the *Spetsnaz GRU* troops, which are attached to the army, are considered to be the best-trained forces of the Russian Federation and among the most formidable in the world. *Spetsnaz FSB* is controlled by the Federal Security Bureau, and *Spetsnaz GRU* specialises in cross-border operations.

During the 1980s *Spetsnaz* troops played a key role in the Soviet invasion and subsequent occupation of Afghanistan, where they became skilled in what has subsequently become known as asymmetric warfare, fighting the Mujahadeen . . . although, as it happens, to no avail.

US SPECIAL FORCES
Delta Force
Motto: Surprise, speed, success!
Based at Fort Bragg in North Carolina, the US Army's Delta Force – 1st Special Forces Operational Detachment-Delta – was formed in 1977 by Colonel Charles Beckwith, and was strongly influenced by the structure and tactics of the British SAS. It is an elite Special Forces unit intended for conducting what the US Department of Defense (DoD) describes as 'tier one counter-terrorism' and special missions, with the emphasis on terrorism, hostage rescue and close-quarters fighting, and is officially referred to by the DoD as the Combat Applications Group. Recruits for Delta Force are drawn primarily from the Rangers and the Green Berets.

Delta Force has been involved in missions in the Gulf War in 1991, in Iraq in 2003 and in Afghanistan.

Green Berets
Motto: De oppresso liber – To liberate the oppressed
Founded in 1952 to act as counter-insurgency troops against Communist aggression, and headquartered at Fort Bragg, Carolina, the US Army Special Forces currently form part of the United States Army Special Operations Command (USASOC). Universally known as the 'Green Berets' as a result of the distinctive headgear that was first adopted unofficially in 1954, the Green Berets proved themselves during the Vietnam War, undertaking various clandestine intelligence-gathering operations, often crossing into Laos and Cambodia in pursuit of an elusive enemy.

Between 1981 and 1985 the Green Berets were involved in training local troops to help the government of El Salvador in its fight against the left-wing FMLN. At the end of the decade the Green Berets, together with the Rangers and Navy Seals, were responsible for Operation Just Cause, the removal of the leader of Panama, Manuel Noriega, who had been allowing the country to be used as a way-station by drug runners. During Operation Desert Shield US Special Forces were involved in training in Kuwait, as well as operating behind enemy lines to provide intelligence regarding the location of Scud missile sites. After the war assistance was given in reconstituting the Kuwaiti armed forces. The Green Berets have also been heavily involved in the wars in Iraq and Afghanistan.

Marine Corps
Motto: Semper fidelis – Forever faithful

Originally founded as an amphibious landing force, and nicknamed the 'Leathernecks', the US Marine Corps can trace its origins back to the American War of Independence, and is modelled on the structure and tactics of the British Royal Marines. Still famed for its fearless amphibious assaults, the Marine Corps uses the mobility of the United States Navy to deliver combined-arms task forces rapidly, and has performed more than 300 landings on foreign shores. 'The Corps', as it is often known, is structured in such a way as to provide what is virtually a self-contained army, and is able to operate completely independently when necessary, under almost any conditions.

Currently the US Marine Corps includes just under 203,000 active duty Marines and some 40,000 reserves, and is larger than the British Army. Expenditure on the Marine Corps accounts for around 6 per cent of the military budget of the USA.

During the First World War the Marines gained a reputation for ferocious fighting during the battle at Belleau Wood, and notable Marine operations during the second global conflict included fierce fighting between Marines and the Imperial Japanese Army at Guadalcanal, Tarawa, Guam, Tinian and Saipan, and the famed capture of the island of Iwo Jima in 1945. Notable post-war operations include the assault on Hue City in 1968 during the Vietnam War, and the battle for Baghdad in 2003, when Marine troops fought against elite units of the Iraqi Republican Guard.

Rangers
Motto: Rangers lead the way!

Although the modern US Army Rangers were not created until 1942, the term 'Ranger' was first used in North America as far back as the early seventeenth century, with the first Ranger company officially commissioned in 1767 during the war against the Spanish. Rangers also fought in the French and Indian Wars, in the American Revolution, the War of 1812 and the American Civil War.

In 1942 General George C. Marshall founded the modern Rangers concept along the lines of the British Army's commando units. The Rangers trained with British commandos in Northern Ireland, and the unit became famous for its actions in North Africa, on the Anzio beachhead, at Omaha Beach during the landing stages of the invasion of Normandy, and in the Pacific. In September 1950 a Rangers training school was established at Fort Benning, Georgia, with the Rangers officially disbanded in 1953. The Rangers were revived during the Vietnam War in 1961, later becoming known as the 1st and 2nd Battalions, 75th Infantry (Ranger) Regiment in 1974. A third battalion was added in 1980 and the total Rangers strength is now somewhere around 2,500 men.

During the post-war years the Rangers have been deployed in Afghanistan, Grenada, Iraq, Korea, Panama and Vietnam.

Navy SEALs
Motto: The only easy day was yesterday; or It pays to be a winner
The US Navy SEALs were formed during the Second World War, some nine months after the attack on Pearl Harbor, as a joint Army–USMC–Navy unit initially described as the Scouts and Raiders. They saw considerable action during the war years and were also active in Korea; in 1962 President Kennedy commissioned the first two of the US Navy's Sea, Air and Land Teams – commonly known as Navy SEALs – for operations against Communist forces in Vietnam. Today the SEALs number some 2,000 men and provide the US Navy's principal special operations force, as well as being the maritime component of the United States Special Operations Command (USSOCOM) and a vital part of the Naval Special Warfare Command (NSWC).

SEALs have been involved in the Vietnam War, in the invasion of Grenada (known as Operation Urgent Fury), during the long Iran–Iraq war (where they protected US shipping in the Gulf from attack by Iranian forces), in the invasion of Panama, in Iraq and in Afghanistan. SEALs have also been used in the fight to combat Somali pirates, operating off the west coast of Africa.

Operating a fleet of Jeeps and Chevrolet modified trucks, the Long Range Desert Group (LRDG) was formed in June 1940 and ranged across the Western Desert and the Mediterranean area during the Second World War. (*Warehouse Collection*)

A stripped and topless Canadian Military Pattern 30cwt 4x4 patrol vehicle (based on the Ford F30 or Chevrolet C30 cargo truck), as might have been used in the Western Desert by the Long Range Desert Group. The angular sand and blue camouflage scheme is typical of vehicles used in the Middle East during the early years of the Second World War. (*Simon Thomson*)

Australian Special Forces preparing to clear a building as part of a demonstration at Holsworthy Barracks as Special Operations Command Australia demonstrated its counter-terrorist capabilities to domestic and overseas delegates attending the Asia Pacific Economic Cooperation (APEC) Australia 2007 Security Conference at Holsworthy on 13 December 2006. Although not the type of vehicle normally associated with Special Forces, the basically civilian sport utility vehicle lends itself well to what is essentially a policing role. (*Leading Aircraftman Rodney Welch; ADF*)

Opposite, top: Replica SAS Jeep armed with five Vickers K machine guns, four of which are carried on twin mounts. Note the jerrycans of fuel and water strapped across the flat top of the bonnet and the personal kit stowed in every possible space. (*Simon Thomson*)

Opposite, bottom: As the SAS Jeeps wore out, the regiment replaced them with similarly modified 86in-wheelbase Series 1 Land Rovers, designated FV18006. The first was converted in April 1955, with two more produced in January 1956 and a further six in February 1957; there were also subsequent conversions of the 88in chassis. (*Simon Thomson*)

The Land Rover-based vehicle that is universally known as the 'Pink Panther' has firmly established the style for almost every subsequent special operations vehicle. Using experience gained on operations with existing vehicles, the military requirements were drawn up by the SAS Regiment in 1964. The first vehicle was accepted into service in October 1968. (*Warehouse Collection*)

Even though most Special Forces vehicles spend their time in the desert, the most modern technology can still be defeated by difficult terrain. These US soldiers have managed to get their HMMWV firmly bogged down in sand in southern Afghanistan. (*US DoD*)

Armed with a 0.50in heavy machine gun in the rear and a 7.62mm general-purpose machine gun (GPMG) on the scuttle, this WMIK-equipped Land Rover Wolf XD Defender was photographed in Kuwait. Note the 'friend-or-foe' identification panel on the roll-cage. (*Warehouse Collection*)

The 34th Commandant of the US Marine Corps, General James T. Conway (kneeling, right), talks to one of his men regarding the light armoured vehicle (LAV) during a visit to Combat Outpost Payne in Afghanistan in August 2010. (*Corporal Erin A. Kirk-Cuomo; US Marine Corps*)

Opposite, top: There is no official definition of what constitutes a 'Special Forces vehicle' and in some situations both motorcycles and quad bikes can be pressed into service. This US Army Ranger is at the wheel of a John Deere Gator, delivering weapons and cargo to Kandahar Airport during Operation Enduring Freedom in 2002. (*Photographer's Mate Ted Banks; US DoD*)

Opposite, bottom: Training forms a vital part of any Special Forces unit, including cross-training with allies and mentoring friendly rebel troops in politically delicate situations. These US Marines, part of 2nd Platoon, Charlie Company, Battalion Landing Team are training alongside men of the French Foreign Legion in Djibouti. The vehicle is a French ACMAT VLRA-2 4x4 light reconnaissance and support vehicle; there is also a long-range patrol version of this vehicle. (*Gunnery Sergeant James Frank; US Marine Corps*)

A Chenowth light strike vehicle (LSV) operating at speed in the climate and terrain for which it was designed. TOW missile launchers are strapped to the roll-cage and the vehicle carries two machine guns. Although successful in their designated role, these vehicles have been superseded by the HMMWV. (*Warehouse Collection*)

Impressed with the performance of the Land Rover Defender during the Gulf War, the US Rangers asked Land Rover to design a version to their specification. Described as the 'special operations vehicle' (SOV), it was seen as a replacement for the ageing M151A2 gun platform. Delivery of sixty vehicles started in 1993, with the vehicle also being made available to other military customers. This is the prototype, which is now in private hands. (*Warehouse Collection*)

The AM General high-mobility multi-purpose wheeled vehicle (HMMWV) is used by all US forces in a variety of roles, including a Special Forces version. Although no lightweight, it offers a very high standard of off-road performance and the modular construction allows easy conversion from one role to another. The lead vehicle in this convoy, photographed struggling through an Iraqi dust storm, is armed with a light machine gun and a Raytheon TOW anti-tank missile launcher. (*Jim Gordon*)

Opposite, top: A Westland Lynx battlefield helicopter makes a rendezvous with British troops in a WMIK-equipped Land Rover Wolf XD Defender. The Lynx AH.7 attack/utility helicopter forms a vital component of the British Commando Helicopter Force (CHF). (*Warehouse Collection*)

Opposite, bottom: US Marines of Charlie Company, 3rd Light Armored Reconnaissance Battalion use their light armoured vehicle (LAV) as a base as they scan the hills of the Bahram Chah valley for an insurgent mortar spotter during Operation Rawhide II in Helmand Province, Afghanistan. (*Sergeant Jeremy Ross; US Marine Corps*)

The Land Rover Defender desert patrol vehicle was used by the British SAS as a replacement for the 'Pink Panther'. The vehicle is armed with a MILAN wire-guided anti-tank missile and a 7.62mm general-purpose machine gun (GPMG). Note the Velcro surrounding the headlights, which allows fabric to be attached to reduce reflections. (*Warehouse Collection*)

Opposite, top: All Special Forces vehicles are required to carry considerable amounts of personal kit, as well as fuel, water, food and spare parts, and in this context the rear stowage basket of the WMIK-equipped Land Rover Wolf XD Defender is very useful. (*Warehouse Collection*)

Opposite, bottom: Sometimes only the heavy metal will do. Here, US Marines provide a security watch during a patrol near Combat Outpost Ouellette in Helmand Province, using an Abrams M1A1A main battle tank as an observation point. (*Sergeant Jesse J. Johnson; US Marine Corps*)

A battered and tired-looking Land Rover Wolf XD Defender makes a rendezvous with a Chinook helicopter. Note the ubiquitous sand channels and the deep-water snorkel on the right-hand front mudguard. (*Warehouse Collection*)

Chapter Three

Anatomy of Special Forces Vehicles

During the Second World War the SAS was equipped with a fleet of much-modified Jeeps. Typically the machines were stripped of all unnecessary items before being stowed with fuel, water, ammunition and personal kit in every available space to allow the vehicles to act as a self-contained base for operations. These Jeeps were replaced by similarly modified Series I Land Rovers in the 1950s and then by the iconic 'Pink Panthers', the Series IIA-based Land Rovers that have effectively established the basic design for the modern Special Forces vehicle.

A second type of Special Forces machine began to emerge in the 1980s, based on, of all things, civilian buggies designed for racing over sand dunes. Described as light strike or fast attack vehicles, these machines were typically designed around a tubular space-frame that also formed an integral roll-cage and often a weapons mount as well, and could be protected by Kevlar composite armour. Power was typically provided by a rear-mounted, often air-cooled, engine, frequently supplied by Volkswagen, driving either the rear axle or both axles, making the machines both fast and agile, particularly in loose sand and light soil. Cross-country performance was enhanced by the light weight combined with features such as heavy-duty independent suspension and limited-slip axles. With a two- or three-man crew, these machines were in their element making forays deep into hostile enemy territory to destroy fuel or ammunition dumps, or to gather intelligence.

These well-armed dune buggy-style machines, such as the Longline Cobra and Chenowth light strike vehicles (LSV), were favoured by both US and British Special Forces for operations in the Middle East during and after the Gulf War. More recently, in operations in Afghanistan, British Special Forces have started to replace their Land Rovers with the larger Supacat Jackal and Coyote vehicles, which, as well as being able to carry larger amounts of kit, also offer a degree of armoured protection. For as long as Special Forces are required to operate in the hostile terrain typical of places such as the Middle East and Afghanistan, it may well be that vehicles such as the Jackal and the Coyote represent the way forward.

THE LAND ROVER 'PINK PANTHER'

The War Office 'User Handbook (Army code 22205, August 1969)' nicely sums up the essence of the 'Pink Panther' and, indeed, all Special Forces vehicles, with these few words: 'the vehicle [is a Rover 11, long-wheelbase 12-volt model] fitted with armament, pyrotechnics, navigation and camping equipment for extended operation away from base'. And, in a nutshell, that is what Special Forces vehicles are all about.

Using the 'Pink Panther' as an example, let's take a look at what equipment is actually involved and carried on, or in, the vehicle.

Forward machine-gun mounting

A number 5 Mk 1 machine-gun mounting, designed to accept a MAG L7 (or later) 7.62mm general-purpose machine gun (GPMG or 'gimpy'), is fitted to the top of a square-section telescopic tubular column, which is bolted through the floor to a chassis cross-member and steadied by being bolted to the top of the dashboard. A compression spring inserted into the base of the tube, acting on blocks between the inner and outer sections, is used to balance the weight of the gun. A captive quick-release pin fitted through the two sections of the tubes allows the two guns to be set at one of two heights, 9 inches apart.

The mounting allows a maximum elevation for the gun of 45 degrees and a depression of 10 degrees, and enables the gun to be locked in any intermediate position by a handwheel-operated clamp. The mount also provides three ranges of traverse via a lift-and-turn knurled knob: in the highest position the gun can be rotated through a full 360 degrees; in the middle position the traverse is 270 degrees, while in the lowest position the traverse is reduced to 120 degrees.

A box for spent cartridge cases is secured to the right-hand side of the mounting, and there is a quick-release attachment for the live ammunition container.

Rear machine-gun mounting

A similar mounting is provided at the rear for a second GPMG.

Ammunition stowage

There are three stowage compartments for 7.62mm machine-gun ammunition, one between the two front seats, and one on each side of the rear compartment, behind the auxiliary fuel tanks and over the wheel arches.

Stowage for Carl Gustav recoilless rifle

Stowage facilities are also provided across the front end of the rear compartment, on top of the decking over the auxiliary fuel tanks, for a Carl Gustav 84mm recoilless rifle – often referred to by British troops as a 'Charlie G'. This very effective weapon first appeared in 1948, and because of its accuracy it was quickly adopted across the West as a primary hand-held anti-tank gun. Ammunition for the rifle is stowed in the rear compartment.

Stowage for self-loading rifle
There is a holster, consisting of a tapered steel stowage compartment, for a standard infantry self-loading rifle (SLR) on the outside surface of each front mudguard.

Grenades
Stowage for grenades is provided at the outer extremities of the dashboard, and at the left-hand side of the vehicle adjacent to the rear gunner's position.

Smoke dischargers
Four sets of three electrically operated smoke dischargers are fitted, one set at each of the front and rear corners of the vehicle. On each set all three tubes are at a fixed elevation of 25 degrees, but each has a different traverse (0, 15 and 30 degrees from the centre-line of the vehicle). The dischargers are fired by four push-button control switches located below the centre of the instrument panel, and the discharger tubes are protected by rubber covers when not in use.

Signal pistol
A standard signal pistol is stowed under the main instrument panel in a split steel tube, with the corresponding cartridges stowed in a steel box located over the left-hand side of the clutch/gearbox cover.

Communications equipment
An open-box mounting is fitted to the centre of the forward face of the rear compartment to accommodate a Larkspur Type A43 portable battery-operated ground-to-air UHF transceiver, with space for a battery charger immediately beneath. Studs are provided on the decking over the left-hand auxiliary fuel tank to accept a Larkspur A123, together with its antenna.

Navigation equipment
Two types of compass are generally carried: a sun compass, designed to be mounted either above the instrument panel or on the decking above the right-hand auxiliary fuel tank, and a Type E2B illuminated compass installed on a glass-fibre mast mounted above the right-hand end of the dashboard. There is also stowage for a Watts Mk 3 theodolite on the top face of the left-hand mudguard, and a pair of

binoculars to the right-hand end of the dashboard. A canvas bag is also provided to allow the delicate navigation equipment to be packed safely out of harm's way.

Camping equipment
The camping equipment consists of a bivouac tent carried in a canvas pannier basket, installed at the aft end of the rear compartment. The basket can be folded up into a vertical position or held in the horizontal position, where it is supported on a hinged platform at the rear by means of chains. There is also stowage space in the rear compartment for a solid-fuel cooker and a metal dixie.

Jerrycan stowage
There is stowage space in the rear pannier for four jerrycans, as well as space for four more in folding tray stowage, two in the front and two in the rear compartment.

Miscellaneous equipment
Two medical packs are carried, a 'fighting-vehicle' type stowed in the front compartment, and a 'troop-type' pack in the rear compartment. In addition, there are three hand-held all-purpose fire extinguishers, one carried on the top of the bulkhead, one on the decking over the left-hand auxiliary fuel tank and one in the rear compartment.

A vehicle camouflage net is carried over the engine cover, secured by straps, and a manilla tow rope is carried in the rear compartment. Two lengths of pressed-steel planking are stowed, one on either side of the rear compartment, and two shovels are carried in a three-compartment stowage on top of the right-hand mudguard, with a pickaxe stowed in two separate parts on top of the right-hand SLR holster.

Auxiliary fuel tanks
In addition to the two main tanks, each of 10 gallons capacity, there are also two auxiliary fuel tanks, each of 40 gallons, fitted one each side of the forward end of the rear compartment and lagged with asbestos cloth; these tanks are covered by plywood decking, hinged at the front to the top of the bulkhead and secured at the rear by a toggle fastener. There are also two expansion tanks, each with a capacity of 5 gallons, giving a total fuel capacity of 110 gallons and a range of around 1,500 miles.

Additional electrical equipment
The vehicle's electrical system is modified to suit the arduous conditions likely to be encountered in service. Additional equipment includes a blackout driving light on the

front right-hand mudguard; a portable searchlight, which can be mounted in one of three positions, either on the dashboard, or on top of the rear compartment above either rear wheel arch; an inter-vehicle starting socket; a special ignition switch, which allows the driver to select either 'normal electrics' or 'ignition only', in the latter case preventing accidental switching on of any lighting equipment; a power supply for the smoke dischargers; a light for the sun compass; and sockets to allow the connection of a battery charger and the radio power supply.

That colour ...

During the desert campaigns of the Second World War the LRDG and the SAS painted their vehicles sand colour; later in the war those vehicles that were operated in northwest Europe were finished with the standard matt green of the period. The same was true during the post-war years: vehicles operated in desert climates were sand-coloured, others were matt green.

When the Series IIA 'Pink Panthers' started to enter service in 1968/69 the SAS was busy fighting the Dhofar Rebellion in southern Oman, and it was found that a pink colour provided the perfect camouflage against the local terrain. The distinctive pink paint, which was not a standard military shade, was apparently achieved by simply mixing white and red oxide, the result being liberally applied across every external surface, often including the tyres! It is said that the vehicles were dubbed 'Pink Panthers' in tribute to the elusive jewel thief in the 1963 Peter Sellers film of that name.

However, not all Pink Panthers were pink; all the vehicles built actually left the factory in the standard military Deep Bronze Green and some probably stayed that way for their entire service lives, while others were painted in standard olive drab with the familiar black 'shadow' camouflage.

Personal weapons

Personal weapons favoured by SAS members over the last two or three decades have included the Browning 9mm high-power pistol in its Mk 2 form, the L96A1 sniper rifle, the Belgian Minimi 5.56mm machine gun and the German Heckler & Koch MP5 9mm sub-machine gun.

Lacking that all-important four-wheel drive, the Chevrolet 1533X2 of the Long Range Desert Group (LRDG) was developed from a standard civilian truck. The rear body sides were extended to allow more kit to be carried, and each vehicle was manned by a crew of three. Despite considerable abuse, the vehicles proved very durable. (*Warehouse Collection*)

At first, transport for the SAS Regiment was provided by the LRDG, but the regiment soon developed its distinctive Jeeps. Stripped of all non-essential equipment, the vehicles were adequately armed and provisioned to allow the teams to operate behind enemy lines for several days at a time, causing mayhem to Rommel's logistics. (*Simon Thomson*)

Compare this period photograph of an SAS Jeep with the preserved machines shown elsewhere. The co-driver has a 0.50in heavy machine gun and there is a Vickers K machine gun in the rear. The jerrycans, used to carry both fuel and water, include British, German and US patterns. (*Tank Museum*)

Starting in 1955, the Land Rover Series I (FV18006) began to appear as a replacement for the SAS Jeeps as these started to become unserviceable. The aluminium-alloy body of the Land Rover probably proved less durable than that of the Jeep, but at least there was a tailgate to help in loading supplies. By this time armament included both the elderly Vickers K machine guns and the ubiquitous Bren gun. (*Warehouse Collection*)

T5160/1

These official portrait views of the Land Rover Series IIA 'Pink Panther' (FV18064), as used by the SAS, show the large amounts of equipment that were carried on these vehicles. Although largely based on the standard military Series IIA chassis, the conversion work, which was undertaken by Marshalls of Cambridge following specifications agreed with the regiment, included a reinforced chassis, hydraulic steering damper, beefed-up suspension and a greatly increased fuel capacity. The first of seventy-two examples entered service in October 1968 and many survived into the 1980s. (*Warehouse Collection*)

T5160/2

The Australian SAS also uses Land Rovers for many operations, but prefers the home-grown Perentie MC2HD 6x6. Constructed on a purpose-designed heavy-duty chassis, the MC2HD is powered by an Isuzu 4BD1 four-cylinder turbocharged diesel engine, driving through an early Range Rover transmission. The rear bogie was developed from work carried out in Britain by SMC Engineering of Bristol, and drive to the rear-most axle comes from a separate propeller shaft on the transfer box power take-off. (*Australian Defence Force*)

While most Special Forces vehicles produced during the three or four decades following the end of the Second World War owed much to the original SAS Jeeps, the US Army, the Marine Corps and the US Navy SEALs started a new trend in the early 1980s by purchasing what was effectively a militarised sand dune buggy. Powered by a rear-mounted Volkswagen air-cooled engine, the Chenowth light strike vehicle was fast and capable, and these vehicles were used effectively during Operation Desert Storm. (*Warehouse Collection*)

A Chenowth light strike vehicle photographed during a demonstration to the press. The vehicle was designed to be operated by a crew of three, with stowage facilities provided along the sides of the body by means of baskets. (*Warehouse Collection*)

Opposite, top: There are times when even the smallest conventional special operations vehicle is too large and during the Gulf War both the British and Australian SAS carried Harley-Davidson MT500 medium motorcycles in the backs of Land Rovers; the same model has also been used by USAF Combat Control Teams. Later models use a document storage pannier to conceal the heat signature of the engine. (*Warehouse Collection*)

Opposite, bottom: Partly based on Land Rover components, the British Esarco 6x6 was put forward unsuccessfully by its makers as a possible contender for the all-terrain mobile platform (ATMP) role; the contract was eventually won by Supacat. This photograph was taken during trials in Singapore in 1988. (*Simon Thomson*)

A US Army up-armoured HMMWV weapons carrier (M1025 or M1043 variant) mounting a light machine gun on a ring mount and armoured shield. The HMMWV – frequently referred to by US troops as the 'Humvee', and also produced in a special desert version dubbed the 'Dumvee' – is produced in a range of variants and is widely used by all US Special Forces. (*Warehouse Collection*)

Opposite, top: Although only two prototypes were constructed and there were no sales, the HMMWV-based Alvis Shadow might be described as the 'missing link', neatly creating the bridge between more traditional Special Forces machines such as the 'Pink Panther' and the heavyweight vehicles which have emerged recently, such as the Supacat Jackal. One of the prototypes is now in private hands. (*Simon Thomson*)

Opposite, bottom: Photographed in 2008, this British Army Jackal is being put through its paces at Camp Bastion, Afghanistan. The Jackal has been used to replace the WMIK-equipped Land Rover Wolf XD Defender; although still open-topped, it provides better protection against ground-based improvised explosive devices. (*Corporal Ian Houlding; MoD Crown copyright*)

Originally dating back to 1965, the backbone chassis and swing-arm suspension of the Austrian-designed Pinzgauer 710 offers exceptional off-road performance. The 'WP' Special Forces variant, which the makers describe as an air-portable weapons platform, was developed in conjunction with Ricardo Special Vehicles and is fitted with Ricardo's WMIK combined roll-cage and weapons mount. (*Warehouse Collection*)

The Supacat HMT 600 Coyote is effectively a 6x6 variant of the Jackal, with which it shares automotive components. The increased carrying capacity means that it is ideal for use as a resupply vehicle in areas which might be inaccessible to more conventional machines. It is being considered for use by the SAS. (*Andrew Linnett; MoD Crown copyright*)

Originally developed by an Austrian company in 1979, the Mercedes-Benz G-Wagen was inspired by proposals for a military vehicle, developed by Shah Mohammad Reza Pahlavi of Iran and first sold on the civilian market. Its excellent off-road performance and reliability mean that it has subsequently found large numbers of military buyers, and both Peugeot and the US company Stewart & Stevenson use the G-Wagen as the basis for a Special Forces vehicle. Described as the 'interim fast attack vehicle' (IFAV), the Stewart & Stevenson version is in service with the US Marine Corps. (*Simon Thomson*)

Although currently being superseded by the Supacat Jackal in many roles, the Land Rover has always been a favourite of British Special Forces since the mid-1950s. This Wolf XD Defender shows the Ricardo WMIK combined roll-cage and weapons mount and the rear stowage basket. The Iraqi boys are obviously admiring the sand channels strapped to the roll-cage, and the 7.62mm general-purpose machine gun (GPMG). (*Warehouse Collection*)

A well-equipped Land Rover Defender desert patrol vehicle (DPV) being off-loaded from a Leyland-Scammell DROPS (demountable rack off-loading and pick-up system) vehicle. The DROPS system allows easy transfer of loads between vehicle types, and thus obviates the need for constant unloading and reloading. (*Simon Thomson*)

A close-up view of a typical, heavily loaded Special Forces vehicle on patrol – in this case a Land Rover Defender. Note the twin machine-gun mounts, exhaust pipe extension coiled up on the engine cover, canvas-covered headlamps and typically be-goggled crew. (*Warehouse Collection*)

Chapter Four

Vehicles Used by Special Forces

It is not strictly accurate to talk of 'the Special Forces vehicle', since there is no single vehicle type that is appropriate to every aspect of the role, and yet, perversely, the correct choice of vehicle can be crucial to the success or failure of a mission. The basic principles established by the SAS and the LRDG still hold good: compare the features of the original SAS Jeep with the latest Land Rover WMIK and you will see that little has actually changed. Most Special Forces vehicles still tend to be lightweight, to allow air-dropping, and are also adapted to carry sufficient firepower and supplies to permit independent operation behind enemy lines for several days. If there is any change it is that modern machines tend to offer more power and reliability compared to the original Jeeps and Land Rovers that were adapted to this role, often combined with enhanced off-road performance.

However, it is interesting to note that almost all Special Forces vehicles are essentially creatures born out of desert warfare. It is equally intriguing to wonder what might happen if peace were to miraculously break out in the Middle East and today's asymmetric conflicts were to be replaced by a more conventional form of warfare. What would be the role of Special Forces if, as would have happened had the Cold War turned hotter, we had similarly sized and similarly equipped conventional armies facing one another? And what types of vehicle might be appropriate in this situation?

The vehicles described here are typical of those used, or intended to be used, by Special Forces over the last seventy years or so, and while the list is by no means exhaustive, it covers most of the major vehicle types as well as a handful that never made it into production.

AB3 BLACK IRIS

Developed by SHP Motorsports and the King Abdullah Design & Development Bureau (KADDB) in Jordan to replace the ageing M151 in service with the Jordanian Special Operations Command, the AB3 is a lightweight Special Forces vehicle constructed on a space-frame chassis, with a front-mounted diesel engine driving the rear axle through a five-speed manual gearbox; a 4×4 version was also planned. Suspension is independent, using helical springs and adjustable gas/hydraulic shock absorbers. Field trials were completed in the year 2000 and around a hundred vehicles were constructed. A civilian version was also mooted.

There are two seats at the front, together with a rear cargo area that can also be adapted to accommodate three men. The rear roll-cage can be used to mount a heavy machine gun.

Designed by the King Abdullah Design & Development Bureau (KADDB) in Jordan, in conjunction with SHP Motorsports, the AB3 Black Iris is a lightweight Special Forces vehicle in service with the Jordanian Special Operations Command, with an initial order for a hundred vehicles. (*Warehouse Collection*)

ALVIS SHADOW SOV

The Alvis Shadow SOV Special Forces vehicle was developed for the British Army but never adopted. It was based on a shortened and narrowed version of the US M1113 extended-capability variant of the high-mobility multi-purpose wheeled vehicle range, the reductions in size enabling two Shadows to be carried in a CH-7 Chinook helicopter, or three in the fuselage of a C-130 Hercules transport aircraft; the vehicle could also be air-dropped on a medium stressed platform.

The GM 6.5-litre V8 diesel engine and four-speed Hydramatic automatic gearbox of the donor vehicle were retained, as were the independently suspended front and rear axles. A stainless-steel roll-cage/weapons mount was added, together with a drop-down tailgate designed to double as a workbench. A front-mounted electric winch was fitted as standard, and customers were encouraged to choose their own weapons systems.

Although only two prototypes were constructed, the Shadow probably pointed the way forward for the larger Special Forces machines that are now starting to enter service. It was displayed at various defence exhibitions, and as of 2001 was still being described as 'available', albeit now being offered as a police and security vehicle. However, none was sold, and the prototypes were eventually disposed of, with at least one being preserved.

Dating from the 1990s, the diesel-powered Alvis Shadow SOV Special Forces vehicle was based on the chassis and automotive equipment of the American HMMWV and was developed for the British Army, although it was not adopted. (*Warehouse Collection*)

The Alvis Shadow was displayed at various defence exhibitions around the world, but when there were no military orders forthcoming, it was also offered as an internal security or police vehicle. There was never any production beyond the two prototypes. (*Warehouse Collection*)

AM GENERAL HMMWV

AM General's high-mobility multi-purpose wheeled vehicle (HMMWV) – better known as the 'Humvee' – has been modified and adapted for the Special Forces role and has been deployed by all the US Army's Special Forces.

Design work on what became the 'Humvee' began in 1978, when the US Army stated that it had a requirement for a go-anywhere, high-mobility, multi-purpose vehicle that could perform the roles of cargo truck, communications vehicle, weapons platform, personnel carrier and command and reconnaissance vehicle. A detailed specification was published in 1980, and the US motor industry was invited to submit prototypes. The Chrysler Corporation, Teledyne Continental and AM General all put development vehicles forward for consideration; the Food Machinery Corporation (FMC) also intended to submit a prototype but was eventually forced to withdraw due to other commitments.

Prototype and pre-production HMMWVs were tested for more than 600,000 miles over rugged courses designed to simulate world-wide off-road conditions in combat environments, with drivers from both the military and the manufacturers doing everything possible to break them. HMMWVs were driven over rocky hills, through deep sand and mud, in water up to 60in deep, in Nevada's desert heat and the Arctic cold, proving themselves to be maintainable, reliable and eminently survivable, even in the most difficult conditions. The AM General vehicle was the first to successfully complete the trials programme and, since it had already shown that it offered the best performance, and was the lightest of the three, it was selected for production. In 1983 AM General was awarded a $1.2 billion contract to produce 55,000 vehicles in fifteen different configurations over a five-year period; the contract also included an option to increase the number of vehicles purchased by 100 per cent during each of the five option years. In its first incarnation the vehicle was officially designated as the M998 Series, and production vehicles started to enter service in late 1983/early 1984, AM General having taken only six months to tool up for production.

The HMMWV is a technologically advanced multi-purpose vehicle which is mobile, versatile and fast, and which provides superior mobility in the field. During initial 'as produced' tests, the new vehicle proved to be nearly twice as durable as the army required. Despite its size, it is genuinely air-transportable, with three HMMWVs deployed in the C-130 Hercules transport, six in the C-141B Starlifter and fifteen in the C-5A Galaxy; under combat conditions it can be delivered by the low-altitude parachute-extraction system (LAPES) without the aircraft having to land. In tactical operations two HMMWVs can be slung from a CH-47 Chinook or a CH-53 helicopter, while one can be slung from a UH-60A Blackhawk.

The original production M998 was powered by a 6.2-litre Detroit-Diesel V8 water-cooled diesel engine producing 150bhp, driving through a three-speed automatic transmission to both axles. The vehicle has full-time four-wheel drive, independent suspension, steep approach and departure angles, 60 per cent gradient climbing ability and a 60in fording capability, giving the HMMWV exceptional off-road abilities. It is also fast, being able to accelerate to 30mph in eight seconds. The payload of the original vehicle was 2500lb but in 1992, in response to a military request for a greater capacity, the M1097 'heavy Hummer variant' was introduced. Although the engine and drive-train remained unchanged, the frame and suspension were reinforced to give a payload increase to 4,400lb.

In 1994 the base level M998 was replaced with the M998A1 series, and the current M998A2 series was introduced in 1995. The new series was fitted with a more powerful 6.5-litre naturally aspirated diesel engine, with the power output upped to 160bhp, electronically controlled four-speed automatic transmission, and a redesigned emissions system meeting the US government's 1995 standard. The 'expanded capacity vehicle' (ECV), also introduced in 1995, had modified differentials, brakes, half-shafts and chassis, one-piece wheels and run-flat tyres, allowing the payload to be increased to 5,100lb. The ill-fated attempt to rescue downed helicopter crews in Somalia in 1993 had also indicated that there was a need for a light armoured utility vehicle and O'Gara-Hess & Eisenhardt used this chassis as the basis for the M1109 and M1114 up-armoured variants.

The HMMWV is a genuine multi-purpose platform and, although there were initially just five base models, these were developed into fifteen different configurations. This meant that, using one set of common parts, the HMMWV chassis was able to fulfil the weapons systems, command and control systems, field ambulance, and ammunition, troop and general cargo transport roles, as well as providing an excellent platform for adaptation to the Special Forces role. Late model M1026 vehicles which have been adapted in US Army workshops for the desert role are sometimes described as the 'Dumvee'.

Up-armoured HMMWV of the US Marine Corps, mounting an M220 tube-launched, optically tracked, wire-guided (TOW) missile and M240 machine gun. The vehicle is assigned to Combined Armored Assault Team 1, Weapons Company, 2nd Battalion, 7th Marine Regiment, and was photographed at Forward Operation Base Delaram, Afghanistan in 2008. (*Lance Corporal Gene Allen Ainsworth III; US Marine Corps*)

Up-armoured (M1025 or M1043 variant) HMMWVs of the US Marine Corps. The vehicles are with Marine Wing Support Squadron 374, attached to a quick-reaction force team, and were photographed during training at Marine Corps Air Station Yuma, Arizona. (*Corporal Benjamin R. Reynolds; US Marine Corps*)

The HMMWV is a genuine multi-purpose platform easily adapted to any number of roles. It is widely used by US Special Forces in a variety of configurations, including this factory-designed HMMWV special operations vehicle. (*Warehouse Collection*)

US Navy HMMWVs photographed in a mocked-up urban environment as SEALs assigned to Seal Team 17 conduct mobility training as part of a pre-deployment work-up cycle in December 2010 on San Clemente Island, California. Navy SEALs provide the maritime component of US Special Operations Forces. (*Mass Communication Specialist 2nd Class Eddie Harrison; US Navy*)

Convoy of US Marine Corps' up-armoured HMMWVs patrolling in Iraq. Note the 'friend or foe' recognition panel on the rear door, typical of vehicles operated in Iraq during Operation Iraqi Freedom. The lead vehicle is the M1114 variant, and the sign on the radiator grille warns civilians against coming too close to the vehicle, which might suggest that they had hostile intent. (*Warehouse Collection*)

AUVERLAND TYPE A3F

With a wheelbase of just 88in, the French Auverland Type A3F fast attack vehicle is sufficiently compact to be carried inside the French NH90 helicopter or to be dropped by parachute. The A3F was developed from the company's A3 light vehicle, and was first shown to the public in June 1994, with two variants available: the VAL is a logistics vehicle with a rear load-bed dimensioned to carry a 1,650lb NATO pallet, while the VAC is a command version, designed to mount a 5.56mm or 7.62mm machine gun on a front mount, together with a 12.7mm heavy machine gun, 40mm grenade launcher, 20mm GIAT cannon or MILAN anti-tank missile launcher on a central mount.

Power comes from the same type of Peugeot XUD 9TF 2-litre turbocharged diesel engine as the base vehicle, driving both axles through a five-speed gearbox and two-speed transfer case, giving a top speed on the road of more than 75mph. Various modifications have been made to the base vehicle to suit the Special Forces role, including changes to the helical coil spring suspension, modified axles and drive-shafts, and a tubular space-frame for the body. Modular composite armour and run-flat tyres can also be specified. A longer wheelbase version has also been trialled by the French Army.

An initial batch of a hundred vehicles was ordered in early 1998, consisting of fifty of each variant; this was followed by a further 154. The A3F is deployed exclusively by the French Army's 11th Airborne Division.

The French-built Auverland A3F fast attack vehicle was designed in the early 1990s; its compact dimensions were intended to allow delivery by helicopter. Around 250 of these vehicles were delivered to the French Army's 11th Airborne Division from early 1998. Modular composite armour can be fitted to enhance crew protection. (*Warehouse Collection*)

CHENOWTH LSV

First appearing in the 1980s, when it was marketed as a two-seat fast attack vehicle (FAV), the US Army's M1040 Chenowth light strike vehicle (LSV) is a military development of the classic dune racing buggy. It is constructed around a tubular space-frame of chrome-molybdenum steel, with a rear-mounted air-cooled petrol engine, and offers high-speed performance across country, combined with low weight and excellent manoeuvrability. Its small size allows it to be easily transported by helicopter or parachute-dropped from a transport aircraft.

Designed for a crew of three – driver, upper gunner/commander and lower gunner – the LSV can be equipped with two 7.62mm machine guns, together with a 12.7mm heavy machine gun or a 40mm grenade launcher, or can carry a wire-guided anti-tank missile, 30mm cannon, anti-tank rocket launcher or surface-to-air missile (SAM). The manufacturer can also incorporate light armour as required. The standard power unit is a 2.2-litre four-cylinder petrol engine, coupled to a four-speed manual gearbox driving the rear wheels only; a diesel engine is also available. There is independent coil-spring suspension at all four wheels and all three crewmembers are provided with full-suspension seats, enabling very fast cross-country speeds of up to 70mph to be maintained.

An upgraded water-cooled diesel-powered advanced light strike vehicle (ALSV) – sometimes described as the Scorpion – was unveiled in October 1996 and was subsequently shown at defence exhibitions in the Middle East. Major features included four-wheel drive and a five-speed gearbox. There was seating for a four-man crew.

Produced by Lockheed-Martin, both the FAV and the LSV were operated by British and US Special Forces during the Gulf War, and the LSV was also used by US Marines and US Navy SEAL units; it has generally been replaced by the HMMWV. Other users include Greece, Mexico, Oman, Portugal and Spain.

Opposite: The M1040 Chenowth light strike vehicle (LSV) is a military development of a civilian dune racing buggy, and features a tubular space-frame, a rear-mounted air-cooled petrol engine and heavy-duty off-road suspension. Prototypes appeared in the 1980s, initially described as fast attack vehicles (FAV). It has seen service with both the US Navy SEALs and the Marine Corps. (*Warehouse Collection*)

Fast and capable across the type of terrain for which it was designed, the Chenowth light strike vehicle (LSV) was first used in anger during Operation Desert Storm; using these LSVs, US Navy SEALs were the first Coalition troops to enter Kuwait City. (*Warehouse Collection*)

The Chenowth LSV has subsequently been replaced by the ubiquitous HMMWV, despite the latter lacking both speed and performance, and a number of the original vehicles have passed into the hands of military vehicle collectors. (*Simon Thomson*)

CHEVROLET 1533X2

When the Long Range Desert Group (LRDG) started operating inside enemy-held territory in Egypt in 1940 it did so using a fleet of modified Chevrolet WA, WB and VA 30cwt 4x2 civilian trucks procured in Alexandria; there were also a number of cut-down Ford C11ADF station wagons. None of these trucks was entirely successful, and from March 1942 the LRDG standardised on the Canadian Chevrolet 1533X2 as a patrol vehicle, eventually acquiring a total of 200 of these vehicles which were heavily modified to the Group's requirements. Each truck was operated by a crew of three or four men.

As deployed by the LRDG, the 1533X2 was a 30cwt civilian truck powered by a six-cylinder overhead-valve petrol engine producing 85bhp from 3,540cc, and driving the rear wheels through a four-speed gearbox and two-speed axle; there were live axles, mounted on semi-elliptical multi-leaf springs, with large-section (10.50–16) sand tyres fitted front and rear. All non-essential items were removed to save weight, including the cab, and the front grille was cut away to improve the flow of air through the radiator; the cooling system was also modified to reduce water loss by including a condenser in a closed circuit. Folding aero-screens were often fitted to the scuttle, and a heavy bumper was fitted at the front, generally incorporating a pusher bar. At the rear the height of the body sides was raised using timber in order to increase the carrying capacity, and radio trucks were fitted with a cabinet to house a British Number 11 radio set.

Each vehicle was fitted with multiple gun mounts, and a machine gun was invariably mounted on a pedestal in the rear. Typical weapons carried included Vickers K light machine guns (actually designed to be mounted on an aircraft), water-cooled Vickers .303in machine guns, Lewis machine guns, Boys anti-tank rifles, Vickers heavy machine guns and American Browning M2 0.50in machine guns. External stowage facilities were provided for fuel and water, personal weapons, ammunition, spare parts for the vehicle, rations, sand channels, personal kit, etc.

The trucks were extremely reliable and were apparently able to withstand considerable abuse without sustaining damage.

Any Special Forces vehicle operating in a desert environment is eventually going to become bogged down in loose sand and will require sand channels to gain traction. The photograph shows a Chevrolet 1533X2 of the Long Range Desert Group. Note the Lewis machine gun, an American-designed weapon dating from the First World War. (*Warehouse Collection*)

The Long Range Desert Group started operations behind enemy lines in 1940 using a fleet of Chevrolet VA, WA and WB 30cwt 4x2 civilian trucks that had been procured in Alexandria. The trucks were modified to suit the role and were often fitted with over-sized sand tyres. (*Warehouse Collection*)

Extensively reworked to suit the arduous demands made on it, the LRDG's Chevrolet 1533X2 was a 30cwt 4x2 truck fitted with a Number 41B welded-steel ammunition body built by Gotfredson. (*Warehouse Collection*)

The front grille of the LRDG Chevrolet was cut away to improve cooling; other modifications included the removal of all non-essential items in order to save weight, the addition of multiple gun mounts, sand tyres, and stowage facilities for fuel and water, personal weapons, ammunition, etc. The vehicles were also fitted with multiple gun mounts. (*Warehouse Collection*)

COBRA LIGHT STRIKE VEHICLE

Developed by the Brighton-based Longline Company, using commercially available components wherever possible, the Cobra light strike vehicle (LSV) was designed as a highly mobile weapons platform for use in hostile environments. The vehicle is constructed around a high-tensile space-frame chassis, which incorporates a tubular roll-cage and a universal weapons mount interface (UMI); there is also a full-length Makrolon polycarbonate protective under-tray. Power comes from a 1.9-litre VAG turbocharged diesel engine installed at the rear, driving through a five-speed gearbox that includes a crawler gear. Both 4x4 and 4x2 drive-lines are available, with lockable differentials or viscous couplings to help enhance off-road performance. The suspension consists of long-travel coil springs and heavy-duty shock absorbers, with double wishbones at the front and trailing arms at the rear. Top speed on the road is 80mph and very high average speeds can be achieved across country. Twin fuel tanks give a maximum range of more than 400 miles, and there are twin radiators and an oil cooler to allow continuous operations in high temperatures.

Designed to be operated by a crew of two, the driver is seated on the left, with the navigator/gunner alongside. The UMI will accept 7.62mm and 12.7mm machine guns, a 40mm grenade launcher, 30mm cannon or MILAN anti-tank missile launcher. A 51mm or 60mm mortar launcher can also be carried.

The Cobra LSV can be fitted into a standard ISO shipping container and can be carried under CH-47 Chinook, Puma and Sea King helicopters, or air-dropped on a medium stressed platform. The vehicle was used by the SAS during the Gulf War where, despite frequent and often serious over-loading, it proved to be an extremely capable machine.

Opposite, top: The British SAS version of the Chenowth LSV was the Cobra. Developed by the Brighton-based Longline Company, it was powered by a 1.9-litre VAG turbocharged diesel engine installed at the rear, driving through a five-speed gearbox. Both two- and four-wheel-drive versions were produced. (*Simon Thomson*)

Opposite, bottom: A privately owned Cobra LSV armed with a 0.50in Browning heavy machine gun on the so-called universal weapons mount interface (UMI) fitted to the roll-cage. The vehicle was used by the SAS during the Gulf War. (*Simon Thomson*)

'Shoot and scoot!' Although there are side stowage baskets, like all buggy-based Special Forces vehicles, the Cobra sacrifices carrying capacity for speed and off-road performance, offering the two-man crew the ability to get in, get the job done and get out ... fast. (*Simon Thomson*)

EMERSON FAST ATTACK VEHICLE

Manufactured by the American Emerson Electric Company, based in St Louis, Missouri, the XM1040 fast attack vehicle (FAV) made its first appearance in 1981, but details are scarce. Period illustrations show a typical space-frame dune buggy-style machine designed for a three- or four-man crew. The standard power unit was a four-cylinder air-cooled petrol engine, but both diesel and turbocharged diesel engines were also available as an option. Mounted at the rear, the engine was coupled to the rear wheels through a four-speed gearbox. A substantial roll-cage could be used as a mount for a 30mm chain gun, 0.50in machine gun, wire-guided missile launcher or 40mm grenade launcher, and a pintle mount could also be fitted in front of the co-driver.

It was rumoured that the US Army had ordered a total of 400 of these vehicles at a price of $80,000 each but that in the end just 125 were constructed, all but ten of which had been scrapped by 1985. Emerson Electric was also said to have been the contractor involved in modifying the Chenowth LSV for military service.

The Emerson Electric fast attack vehicle (FAV) made its first appearance in 1981, and it was rumoured that the US Army had ordered a total of 400 at a price of $80,000 each; in the end just 125 were constructed. Emerson Electric was also involved in helping to develop the Chenowth LSV. (*Warehouse Collection*)

In the style of such machines, the Emerson FAV was powered by a rear-mounted four-cylinder air-cooled petrol engine, but both diesel and turbocharged diesel engines were also available as an option. (*Warehouse Collection*)

The Emerson was fast and agile, and the roll-cage could be used as a mount for a 30mm chain gun, 0.50in machine gun, wire-guided missile launcher or 40mm grenade launcher. (*Warehouse Collection*)

ENGESA EE-VAR

Constructed by the Brazilian ENGESA (Engenheiros Especializados) company, the EE-VAR is a two-man lightweight dune buggy-style machine designed for forward reconnaissance operations and for use as a weapons platform. Like most machines of this type, it is powered by a 1,600cc Volkswagen four-cylinder air-cooled engine, producing 70bhp at 4,000rpm, mounted at the rear, and arranged to drive the rear wheels through a four-speed manual gearbox. There is independent suspension all-round using torsion bars and hydraulic telescopic shock absorbers, with two shock absorbers fitted at each rear wheel station, and the top speed is in the order of 85mph. The floor panels are protected by Kevlar composite armour, and Kevlar panels are also fitted to protect the engine mount at the rear and in front of the driving compartment.

Typical on-board weapons can include 5.56mm, 7.62mm or 12.7mm machine guns, a 40mm grenade launcher and a wire-guided anti-tank missile system.

Opposite, top: Dating from the mid-1980s, the Brazilian ENGESA EE-VAR is a two-man lightweight dune buggy-style machine intended for forward reconnaissance operations and for use as a weapons platform. (*Warehouse Collection*)

Opposite, bottom: The ENGESA EE-VAR is powered by a rear-mounted Volkswagen 1,600cc four-cylinder air-cooled engine, driving the rear wheels through a four-speed gearbox. Weighing little more than 1250lb, the vehicle is capable of a top speed of 85mph. (*Warehouse Collection*)

HARLEY-DAVIDSON MT350, MT500

Originally badged Armstrong, the Harley-Davidson MT350 and MT500 motorcycles were derived from a machine that had been developed in the early 1980s by the Italian SWM company. The rights to the model were sold to Armstrong, with the company spending two years developing the Italian machine into a reliable military motorcycle, beating off competition for a British Ministry of Defence (MoD) contract from sixteen competing manufacturers; in 1985 it was announced that 2,300 MT500s were to be produced for the British Ministry of Defence. Harley-Davidson became involved in 1987.

The MT500 was powered by a Rotax four-valve single-cylinder motor of 485cc, driving through a five-speed transmission. Main features included heavy-duty hydraulic suspension at front and rear, solid-state ignition system, sealed-for-life drive chain with O-ring seals, water- and dust-resistant brakes and aluminium wheel rims. Although the MT500 was generally equipped with a kick-starter, a number were also fitted with electric-start and supplied to the Jordanian and Canadian armies, the Canadians confusingly designating theirs as M50; the model has also been used by USAF Combat Control Teams.

In 1993 the MoD ordered a further 1,570 of these motorcycles from Harley-Davidson, this time specifying the smaller-engined (348cc) MT350, with document boxes fitted to the frame down-tube to 'obscure' the engine and thus conceal the thermal signature.

During the Gulf War both the British and Australian SAS carried these motorcycles in their Land Rovers.

Opposite: Although unarmed and lacking stowage facilities, motorcycles can be used to provide Special Forces with the elements of speed and surprise. The British Army's Armstrong MT350 off-road motorcycle, and the larger MT500, was based on an Italian machine originally designed by SWM and later manufactured by Harley-Davidson. (*Warehouse Collection*)

The Harley-Davidson MT350 was used by both the British and Australian SAS, as well as by Combat Control Teams of the US Air Force. (*Warehouse Collection*)

HOTSPUR SANDRINGHAM S6, S6E

The Hotspur Sandringham was a Land Rover conversion, developed as a private venture by Hotspur Armoured Products Ltd of Bristol and derived from the Sandringham 6, a civilian conversion of the Stage One. Fully approved by Land Rover, the first Hotspur prototype was displayed at the British Army Equipment Exhibition in June 1980, and by early 1981 the prototype had successfully completed a trials programme. Production started in 1982.

The Sandringham was based on a welded hull using Hotspur super-hard opaque steel armour, and the vehicle was originally designed as an armoured personnel carrier; there were two wheelbase lengths, the S6 measured up at 125in, while the S6E was extended to 139in. There was space in the boxy hull for a crew of two in the cab, together with eight fully equipped troops in the rear compartment, seated on longitudinal benches along either side. The hull was designed to provide protection from shell splinters and high-velocity rifle fire (typically NATO 7.62mm ball round) at a minimum 40m distance, and there were six firing ports, with vision blocks. The floor was protected against mine fragments and grenades, and armour was also applied to the engine compartment and radiator, and the wheels were equipped with run-flat bands allowing limited operation on a punctured tyre. Access to the cab was provided by hinged doors with large vision windows of composite glass and plastic construction, with separate polycarbonate anti-spall screens; hinged armoured screens could also be specified for the windscreen and side windows. There was a pair of hinged doors at the rear, with fold-down steps.

Standard equipment included a fan ventilation system, interior lights and a long-range 80-litre explosion-proof fuel tank with a locking cap. Customers could also opt for an engine fire-extinguishing system, internal communication system, barricade ram, smoke or CS gas dischargers, power winch, heavy-duty alternator, diesel engine and a 6x4 drive-line.

Alongside the basic APC, other body variants included a command and communications vehicle, field ambulance, long-distance Special Forces or patrol vehicle, and fuel or water tanker. A commander's cupola could be installed in the roof at the rear, designed to mount a general-purpose machine gun (GPMG).

Another one that didn't make it! The Hotspur Sandringham S6 and the longer-wheelbase S6E were developed as a private venture by Hotspur Armoured Products and based largely on Land Rover components. Among the several variants proposed, there was a long-distance Special Forces or patrol vehicle. (*Warehouse Collection*)

JEEP
SAS Jeeps

By early 1942 the regimental strength of the SAS was up to 130 men; now equipped with twenty Bedford 3-ton trucks and sixteen Jeeps, the unit was sufficiently large to no longer need to rely on the Long Range Desert Group (LRDG) for transport. At the time one of the standard operational tactics of the SAS was to infiltrate a small, lightly armed reconnaissance group into place carrying radio equipment, to be subsequently reinforced by additional, more heavily armed, troops usually travelling in specially equipped Jeeps. Stripped of all non-essential equipment, and bristling with heavy automatic weapons, these Jeeps were modified to provide a well-equipped and well-armed patrol vehicle, capable of carrying two or three men, together with sufficient fuel, rations, ammunition and supplies for extended missions deep into enemy-held territory. Curiously, there does not seem to have been a standard set of modifications and examination of period photographs shows that, although there were features common to all the SAS Jeeps, essentially each appears to have been modified according to the needs of the mission, and with the over-riding intention of reducing superfluous weight.

The vehicles were almost invariably heavily armed: the standard equipment seems to have been a pair of twin-mounted Vickers K .303 observer's machine guns on a pintle ahead of the front passenger's seat – this was originally an aircraft-mounted gun and, with a rate of fire of more than 3,000 rounds per minute from a drum magazine, it was a formidable weapon, offering twice the hitting power of the Bren. There was often a third Vickers, or a .303 Bren gun, on a pedestal mount to the left of the driving position, and a standard infantry-issue water-cooled Lewis machine gun was sometimes carried for use in static firing. Other variations included the use of an M2 0.50in heavy machine gun ahead of the passenger seat, with the twin Vickers units relegated to the rear area ahead of the back seat; other examples show a 0.50in machine gun at the rear. Other weapons were carried to suit the particular mission. The normal 2-inch and 3-inch mortars usually proved useful for destroying enemy targets, as did the PIAT (projectile, infantry, anti-tank) gun. Most SAS raiding parties would also have carried a plentiful supply of number 36 Mills bomb grenades, plus other grenades such as the number 69 Bakelite grenade and the Gammon anti-tank bomb.

The most distinctive of the modifications to the vehicle itself included ad-hoc 'improvements' to the cooling system where, for desert operations, most of the bars of the front grille were cut away to provide optimum airflow through the radiator and thus ensure maximum cooling efficiency. Whether or not it was necessary, this seems to have become something of an SAS trademark, and even Jeeps operating in northwest Europe were normally seen with the distinctive cut-away grille. The open, pressurised cooling system of the standard Jeep was modified to a sealed system using a version of the desert cooling modification kit. A small cylindrical expansion tank was fitted at the front and connected to the radiator overflow via a small pipe, and the system was sealed in such a way that water was allowed to expand into this tank as the engine heated up, but could be drawn back into the main system via the same pipe when the water cooled down and contracted – a process that has subsequently become standard on all motor vehicles.

In the interests of maintaining a low profile, the standard windscreen, hood and hood frame were generally removed and discarded altogether. Even in colder latitudes, the standard windscreen was not fitted, but on some examples heavy bullet-proof glass shields were provided for the front-seat gunner and occasionally the driver, as part of the gun mount. Occasionally the front bumper was also discarded or cut back in the style of the airborne Jeeps in an effort to save more weight. Some vehicles were protected underneath using armour plate, so as to reduce the effects of mine blasts.

A large amount of the available storage space was used to carry fuel, water or ammunition – even the bonnet was pressed into service, often with four jerrycans strapped across the flat surface. Some vehicles also carried additional fuel tanks over the wheel arches in the rear, 'borrowed' from a standard 3-ton truck. During some operations certain Jeeps were assigned to the support role, and armaments were omitted in favour of additional jerrycans – since the Jeeps were relatively fast compared to other trucks of the period, the use of Jeeps in the supply role meant that the operation was not held back by the presence of slower vehicles.

All things considered, the SAS Jeep made a formidable battle wagon and it is hardly surprising that the vehicle effectively became the role model for Land Rover 'Pink Panthers' and today's long-range Special Forces Defenders.

Popski's Private Army Jeeps

Modified Jeeps were also deployed by Popski's Private Army, operating in patrols consisting of six vehicles and sixteen men. The vehicles were stripped of non-essential items, including the windscreen and top, and, as with the LRDG Chevrolets and the SAS Jeeps, most of the radiator grille bars were removed to increase the flow of cooling air through the radiator. Water condensers were also fitted to the radiator so that any water that boiled off was not lost. At the front the standard military bar-grip tyres were generally replaced by road tyres, since these were less likely to break through the crust that forms on desert sand. Armaments included Vickers K or Browning 0.30in and 0.50in machine guns, sometimes on a twin mount, together with a smoke generator. Racks were fitted to carry twelve 4-gallon petrol cans, giving the vehicles a range of between 600 and 700 miles.

At least one of Popski's Jeeps was experimentally fitted with flame-thrower equipment taken from a Canadian Wasp carrier; during trials the equipment apparently singed the eyebrows of the operator and it is believed that it was never used in action.

Needing little introduction, the iconic SAS Jeep is among the most recognisable Special Forces vehicles of all time. The official caption describes this as a 'patrol Jeep fitted with single and twin Vickers K machine guns, and a 0.50in heavy machine gun . . . each vehicle carried at least twenty 4-gallon cans of petrol in addition to food and water for a month'. (*IWM, NA675*)

The SAS Jeep forms a popular subject for military re-enactors and although considerable care is taken to ensure that appearance of the vehicle and the type of kit carried remain authentic, the hard life to which these vehicles were originally subjected means that it is unlikely that originals have survived. (*Simon Thomson*)

Although the official caption describes this well-loaded vehicle as a 'Jeep car' of the Long Range Desert Group (LRDG), the crew are actually wearing SAS badges on their caps. The photograph was taken in Cyrenaica, Libya, in 1942. (*IWM, E20084*)

Popski's Private Army also used specially modified and equipped Jeeps. This replica vehicle displays on the radiator grille the astrolabe unit badge, based on a navigational aid that measured the altitude of stars. Popski's men operated in patrols consisting of six vehicles and sixteen men. (*Simon Thomson*)

LAND ROVER
Series I SAS vehicles

Officially described as 'truck, ¼ ton, 4x4, SAS, Rover Mk 3; FV18006', the first post-war SAS patrol vehicle was basically a modified 86in Series I cargo vehicle. There were few mechanical changes, with little more than the use of higher-rate springs to enable the additional load to be carried. The doors and windscreen were removed, there was no canvas top or frame, the centre section of the front seat was removed, and the passenger seat was modified and repositioned. A radio was fitted in the rear, and there was a large auxiliary fuel tank inside the body, beneath a single rear-facing seat for the radio operator/rear gunner. Considerable additional kit was carried, including jerrycans for fuel and water, stowed on the reinforced front bumper and inside the vehicle, and additional stowage bins and lockers were also fitted wherever practicable. The spare wheel was repositioned to a bracket on the front bumper.

Standard armament included a pair of 7.62mm general-purpose machine guns (GPMG) on a coupled mount at the front, where they could be operated by the co-driver, and a Browning 0.30in machine gun at the rear. A third GPMG was normally stowed beside the driver.

The first vehicle was converted in April 1955; two more were produced in January 1956 and a further six in February 1957. Automotive trials on the original conversion were conducted at the Fighting Vehicle Research & Development Establishment (FVRDE) in late 1957, before the vehicle was returned to the War Office for air-dropping trials. A number of 88in Series I Land Rovers were also converted to this role in a similar manner, remaining in service until 1967, when the Series I was replaced by the long-wheelbase Series IIA – the iconic 'Pink Panther'.

SIIA 'Pink Panther'

The origin of all Special Forces' vehicles can be traced back to the SAS Jeeps and to the Chevrolets used by the LRDG. These, in turn, led to the development of the SAS Series I Land Rovers which replaced the Jeeps in the early 1950s and remained in service until the late 1960s, when it was decided that they should be replaced by a long-wheelbase Land Rover that could provide additional stowage facilities and at the same time allow the operating range to be increased. The vehicle that was developed is universally known as the 'Pink Panther' and it has firmly established the style for almost every subsequent special operations vehicle.

The military requirements for what was officially known as 'truck, GS, SAS, ¾ ton, 4x4, Rover 11, FV18064' – but which everyone now refers to as the 'Pink Panther' – were drawn up by the SAS Regiment in 1964 and were derived from experience gained on operations with the existing vehicles. The vehicle had to be able to carry large quantities of fuel, water and other supplies, and radio communications equipment, as well as a three-man crew, along with sufficient weaponry for attack and self-defence. During the early 1960s the regiment workshops produced a total of twenty-seven prototypes in conjunction with REME, but it was clear that the generally unmodified Series II chassis was not really up to the task. Final development of the vehicle was eventually turned over to FVRDE, with the regiment specifying the equipment which was to be carried. In December 1965 FVRDE issued a new statement of requirements, having inspected the existing vehicles and discussed with the regiment what was required.

The unit-modified SAS vehicles had shown that heavy-duty springs and one-piece 6Lx15in wheels, mounting 9.00–15 sand tyres, could help to overcome failures in these areas as well as increasing mobility, and these modifications were carried forward to the FVRDE prototype. Other changes included the use of twin asbestos-wrapped long-range fuel tanks, which pushed the average operating range to 1,500 miles, and the inclusion of various weapons mounts – the smoke-grenade launchers and machine-gun mounts being developed especially for this role. The FVRDE prototype also sported the familiar body modifications that are now associated with the production vehicles. Although there was some rearrangement of components under the bonnet, the standard 2,286cc four-cylinder Rover petrol engine remained in place, as did the four-speed transmission, and the vehicle also retained the 12V electrical system.

By May 1967 the specification had been finalised and Marshalls (Engineering) of Cambridge was contracted to construct seventy-two examples, based on the Series IIA that had replaced the earlier vehicle in production. Chassis modifications included the replacement of the standard leaf-spring suspension with heavy-duty springs and shock absorbers, and the addition of a hydraulic steering damper. The chassis itself was reinforced at critical stress points, and there were welded guards on the differential housings. The passenger seat was fitted to a raised platform, and additional seats were fitted over rear wheel arches. The range was extended by the use of twin auxiliary fuel tanks increasing the fuel load to 100 gallons. Modifications to the body included removal of the doors, windscreen, hood frame, hood and sills; the spare wheel was moved to a near-horizontal position on the front bumper, and a folding pannier was fitted at the rear.

Aside from the obvious modifications, the vehicles bristled with armaments and bolt-on equipment. Standard weapons issue included two 7.62mm general-purpose machine guns (GPMG), a Carl Gustav 84mm recoilless anti-tank gun and four sets of three electrically fired smoke dischargers. Other essential equipment included a pair of Larkspur morse and ground-to-air radio sets, sun compass, theodolite and standard magnetic compass. Stowage facilities were provided for the crew's self-loading rifles (SLR) in sheet-metal holsters on the front wing side panels; ammunition and grenade stowage facilities were placed between the front seats and on both sides of the rear compartment; a machete was stowed behind the rear seat, plus first-aid facilities, fire extinguishers and water container racks inside the body. Sand channels and pioneer tools were carried on brackets on the outside of the body. A fully loaded 'Pinkie' weighed almost 1,000lb more than the equivalent standard Rover 11.

The first vehicle was ready for inspection in August 1968, and a number of minor issues arose that had to be rectified before the vehicle could be accepted for delivery. On 2 October 1968 the first vehicles were delivered to the regiment, with the remainder delivered during 1969.

The biggest problem in service appears to have been half-shaft failure, and spare shafts were generally carried, but the vehicles were generally considered reliable and were more than capable of carrying out the job for which they had been constructed, with notable operations in Belize, Kenya, Northern Ireland and Oman. The standard operating procedure was developed in Oman, where the vehicles usually went out on patrol in threes; the lead and rear vehicles would be equipped for the attack or reconnaissance role, while the middle one would carry mechanical spares such as starter motors, generators, coils, etc. The distinctive pink colour also first appeared on vehicles used in the Oman Dhofar operation.

Despite operational losses, the vehicles enjoyed a near twenty-year life, finally being replaced by Defender 110-based special operations vehicles (SOV) during the early 1990s.

Desert patrol vehicle

Although numbers of long-wheelbase Series III Land Rovers were converted for long-range patrol work, the conventional leaf-spring suspension often made the going tough and in the late 1980s the first of a series of heavily modified Defender 110 desert patrol vehicles (DPV) started to enter British Army service. Modified by Marshalls of Cambridge, the coil-sprung DPV superficially resembled the 'Pink Panther', lacking doors, a windscreen, top and frame. The suspension was uprated and it was powered by Rover's Buick-derived 3.5-litre V8 petrol engine.

Stowage capacity was increased by virtue of the use of the distinctive body of the civilian high-capacity pick-up truck. An external roll-cage was fitted behind the front seats and there were additional fuel tanks, as well as generous stowage capacity for additional petrol and water jerrycans, ammunition and personal kit. A pair of 7.62mm GPMGs were mounted in the rear, with a third pedestal-mounted on the scuttle; there were also bumper-mounted smoke dischargers.

Special operations vehicle

The Land Rover 'special operations vehicle' (SOV) was originally designed to provide the US Army Rangers with a rapid-reaction, air-portable, all-terrain weapons platform. Impressed by the performance of the British Army's Defenders during the Gulf War, the Rangers saw the SOV as a replacement for the ageing M151A2 gun platform, and in 1991 Land Rover was approached with an outline specification. Like the DPV, the vehicle that Land Rover produced was based on the long-wheelbase four-door Defender 110; it was also offered to other defence customers.

The Rangers SOV, or RSOV, was first seen at the 1992 Eurosatory Show in Paris, with the Rangers taking delivery of sixty examples in 1993. It was powered by the 300 Tdi four-cylinder 2.5-litre turbocharged diesel engine, but a 3.5-litre V8 petrol engine was also offered. The permanent four-wheel drive system, transmission and long-travel coil-spring suspension were based on standard Defender components, and there was a choice of either a 24V 90Ah or 12V 65Ah electrical system, with optional radio suppression. There were no doors, windscreen or hood, but there were additional stowage facilities for ammunition, fuel, water, and other tools and equipment.

For the first time there was a combined roll-bar and weapons mount, and suggested armaments included a pair of 30mm ASP-30 machine guns mounted in tandem at the rear, together with a third machine gun on the scuttle alongside the driver. The vehicle could also mount the Mk 19 40mm mortar launcher, 50mm or 81mm mortar, 0.50in heavy machine gun, AT-4 Stinger, or Carl Gustav 84mm recoilless rifle, with the latter often described as the 'Ranger anti-armor weapon system' (RAAWS). For the personnel carrier role Land Rover claimed that the crew capacity was six men, but as a special operations vehicle it was operated by a crew of three – driver, gunner and leader/gunner.

The SOV was designed to be air-portable in either a C-130 Hercules or in CH-47 Chinook and EH-101 helicopters; it could also be sling-loaded underneath a Chinook, Blackhawk, Sea King or Puma, or para-dropped on a suitable platform.

Rapid deployment vehicle

In 1993 Land Rover exhibited the 'multi-role combat vehicle' (MRCV) at the British Army Equipment Exhibition (BAEE) at Aldershot. Developed partly in conjunction with Longline and Ricardo Special Vehicles, the MRCV was originally shown on the short-wheelbase Defender 90 chassis, but users could also choose the long-wheelbase Defender 110 or 130 as a basis. The MRCV was subsequently renamed the 'rapid deployment vehicle' (RDV) and the option of using the Defender 130 chassis was discontinued.

Two RDV versions were produced. One was a three-seater vehicle configured with a low-profile rear body and superstructure to allow stowage inside a CH-47 or CH-53E helicopter, without the need for dismantling; the other was fitted with a normal-height superstructure to accommodate an additional rear-facing weapons post and gunner.

The vehicle was intended to act either as a weapons platform or as a reconnaissance vehicle, and the modular construction of the rear body enabled easy conversion to one of seven distinct roles, using just hand tools: pedestal mount, MILAN anti-tank platform, multi-purpose ring mount, and personnel or cargo carrier. In the weapons platform role, a 360-degree ring mount was mounted on the roll-bar, which could be used to equip the vehicle with a 40mm grenade launcher, GIAT 20mm cannon, L1A1 0.50in Browning heavy machine gun or 7.62mm GPMG. Pedestal mounts could also be fitted to the scuttle or to the rear for an 0.50in Browning, 40mm grenade launcher or GIAT 20mm cannon, and a 7.62mm GPMG could be mounted anywhere on the rear roll-cage. To enhance stability when used as a gun mount, a lock-out system could be fitted to the suspension. In the MILAN anti-tank (ATGW) configuration, the vehicle carried a complement of six missiles.

The rear cargo area could accept a palletised load and was provided with tie-down points, and a large rear bustle was fitted to provide additional stowage facilities. Like the SOV, there was a choice of electrical systems, with users able to choose between a 65Ah 12V system, or single or twin 24V 50Ah alternators; optional radio suppression facilities were available. None of the modifications made to the vehicle compromised its ability to carry out general service duties when not required for operational duties.

The RDV also offered users the same para-dropping and air-portability options as the SOV.

Weapons mount installation kit (WMIK)

The development of the 'weapons mount installation kit' (WMIK – pronounced 'wimmick') can be traced back to a feature of the Longline light strike vehicle, later known as the Cobra, of 1989, which was favoured by the SAS during the first Gulf War. In 1993 Longline was acquired by the Shoreham-based engineering company Ricardo Special Vehicles, and a year later Ricardo and Land Rover agreed on a partnership arrangement for commercial exploitation of the equipment. While Land Rover started to market the WMIK-equipped MRCV as a standard variation of the so-called 'core' Defender range, Ricardo continued development work on the concept.

In early 1998 the British Army took delivery of a WMIK prototype that could be retro-fitted to the Wolf XD Defender or to any other Land Rover that had already been upgraded by the installation of reinforced outriggers and roll-cage mounts. A trials programme kicked off in March of that year and in 1999 the MoD issued a contract for 135 kits to modify production Land Rovers supplied from Solihull. Within three weeks of the delivery of the first kits, modified vehicles were in service in Sierra Leone.

As with the RDV, the WMIK was designed to allow the base vehicle to be able to act either as a weapons platform or as a reconnaissance vehicle and to be readily adapted from one role to another. Recently the MoD extended the capability of the weapons platform when they announced the purchase of forty new belt-fed automatic lightweight grenade launchers (ALGL) made by Heckler & Koch (HK GMG) for Afghanistan WMIKs. The ALGL can fire up to 360 grenades per minute, lobbing the projectiles at distances of up to 1,750 yards.

Ricardo's original kit was described as the O-WMIK and this formed the basis of the first deliveries to the MoD, although the original DERA swing-arm gun mount was replaced by Ricardo's slewing ring in spring 2002. In mid-2005 the kit was upgraded to become the E-WMIK, which included belly protection plates designed to resist mine blast, and from the beginning of 2006 also allowed the gross vehicle weight to be increased from 3.35 to 3.5 tons. The current designation is the R-WMIK and R-WMIK+; both have additional composite armour in the form of a tub fitted directly to the chassis over which the original body panels are refitted, while the R-WMIK+ also includes automotive upgrades, including a new transmission.

Since 1999 the British Army has put around 300 WMIK-equipped Land Rovers into service, and in 2007 Ricardo announced that a fleet of 200 of the army's WMIK-equipped Land Rover Defenders were being returned for a major upgrade, involving significant enhancements to crew-protection systems as well as an increase in vehicle payload, enabled by re-engineering the chassis and suspension systems. Alongside Land Rover Defender and Wolf XD Defender platforms, the WMIK concept has also been applied to the Supacat 4x4 and 6x6, the Ford F350 to produce the special reconnaissance vehicle (SRV), and the Pinzgauer 6x6. Ricardo has supplied vehicles or kits to the Irish Defence Force, and to the armies of Lithuania, the Netherlands, New Zealand and Sweden. The British Army recently announced the purchase of 130 Jackals, the Supacat-based M-WIMK, as a partial replacement for the Land Rovers.

Ricardo has also produced a range of optional modules and products to supplement the WMIK, many of which can be used with vehicle types other than the Land Rover. These include infra-red lighting systems; lightweight high-back seating systems; wider alloy wheels, adopted as a result of experience in Afghanistan; snorkel; infra-red lighting and a hybrid 12/24V electrical system.

By the mid-1950s the SAS Jeeps were well beyond economic repair and in April 1955 the regiment started to take delivery of replacement vehicles based on a modified short-wheelbase Land Rover Series I (FV18006). At the front is the familiar twin-mount for a pair of Vickers K machine guns, while the rear gun is a 0.50in Browning heavy machine gun; a US Army 3.5in recoilless anti-tank rocket launcher is stowed in the rear compartment. (*Warehouse Collection*)

A replica SAS Series I Land Rover, complete with authentic equipment. The military registration number (66BS14) is correct for the series of SAS vehicles and suggests that this is an 88in wheelbase chassis dating from 1957. (*Warehouse Collection*)

The SAS Land Rovers were stripped of doors, windscreen, canvas top and support and were fitted with higher-rate springs to enable the additional payload to be carried. A single seat in the rear allowed the third crew member to operate the radio or a rear-facing machine gun; an auxiliary fuel tank was fitted under the rear seat. (*Tank Museum*)

Rear view of the SAS Series I Land Rover showing the rear-facing 0.30in Browning machine gun and forward-facing Vickers K guns on a tandem mount. The box visible between the rear gun and the rear seat is a radio set. (*Warehouse Collection*)

Land Rover Series I showing how the spare wheel was relocated to the front of the vehicle to make space for more kit in the rear. Note the machete carried in a steel scabbard on the side of the box supporting the co-driver's seat. (*Warehouse Collection*)

Although generally dubbed the 'Pink Panther' on account of its finish, in fact not all of these Series IIA SAS Land Rovers (FV18064) were actually pink . . . as this well-stowed, Deep Bronze Green painted example demonstrates! They were designed by the regiment and produced by Marshalls of Cambridge, and deliveries started in late 1968. (*Warehouse Collection*)

Rear view of the 'Pink Panther' showing the folding pannier. This example has a 7.62mm general-purpose machine gun on a scuttle mount ahead of the co-driver and there are three-barrel smoke grenade launchers on the rear corners. All 'Pinkies' were fitted with the oversized wheels and tyres of the civilian Series IIB forward-control Land Rover. (*Warehouse Collection*)

An authentic 'Pink Panther' represents the Holy Grail for many military vehicle collectors. Where the real thing is not available, many enthusiasts have constructed replicas, taking considerable care to ensure that the vehicle and the on-board equipment are correct in every respect. (*Simon Thomson*)

Photographed during the Gulf War in 1990, this Series III Land Rover special operations vehicle has been stripped of its doors, windscreen and top, and is wearing oversized front wheels and tyres. Most of the special operations vehicles used during the Gulf War were based on the coil-sprung long-wheelbase Defender. (*IWM, GLF1383*)

A line-up of tired, but still very effective, SAS desert patrol vehicles (DPV). Dating from the mid-1980s, the DPV was based on the 110in Defender chassis and resembled the 'Pink Panther'. However, the use of Rover's Buick-derived 3.5-litre V8 petrol engine meant that this was an altogether more capable machine. (*Roland Groom, Tank Museum*)

The DPV was fitted with the rear body of the civilian high-capacity pick-up truck. Note the launcher for the MILAN wire-guided anti-tank missile mounted in the rear. (*Warehouse Collection*)

A pair of heavily laden, privately owned Land Rover Defender desert patrol vehicles (DPV) showing the amount of kit that could often be required on a mission. (*Simon Thomson*)

Painted all-over matt black, this US Army special operations vehicle (SOV) belongs to B Company, 1st Battalion, 75th Ranger Regiment. A 0.50in heavy machine gun is mounted on the roll-cage, and an M136 light anti-tank weapon is on the ground beside the vehicle. (*Michael Lemke*)

Dating from the early 1990s, the Land Rover 'special operations vehicle' (SOV) was designed for US Army Rangers, who had apparently been impressed by the performance of the British Army's Defenders during the Gulf War. The SOV was seen as a rapid-reaction, air-portable, all-terrain weapons platform and replaced the obsolete M151A2 gun platform. (*Tank Museum*)

The prototype of the Rangers special operations vehicle (SOV) is now in private hands in Britain and is seen regularly at military vehicle shows. Note the combined roll-cage and weapons mount, and the front-mounted electric winch designed for self-recovery. The rings at the vehicle corners are designed to enable it to be lifted by helicopter. (*Simon Thomson*)

Prototype of the Land Rover multi-role combat vehicle that was first shown in 1993. Developed partly in conjunction with Longline and Ricardo Special Vehicles, the vehicle featured the combined roll-cage and weapons mounts also seen on the SOV, and was subsequently renamed the 'rapid deployment vehicle' (RDV). (*Tank Museum*)

Opposite, top: The RDV was produced in two versions, first as a three-seater vehicle with a low-profile rear body and superstructure to allow stowage inside a CH-47 or CH-53E helicopter; secondly, with a normal-height superstructure to accommodate an additional rear-facing weapons post and gunner. (*Warehouse Collection*)

Opposite, bottom: A Land Rover rapid deployment vehicle showing what it is capable of . . . while most Special Forces vehicles seem to have been designed for desert conditions, this RDV shows that it is equally at home in somewhat wetter terrain. (*Warehouse Collection*)

The weapons mount installation kit (WMIK) first appeared on the Cobra light strike vehicle and was developed by Ricardo Special Vehicles as a means of combining a roll-cage with a versatile weapons mount. In 1999 the MoD issued a contract for 135 kits to modify production Land Rovers. (*Warehouse Collection*)

Land Rover Wolf XD Defender patrol vehicle with the WMIK and rear stowage basket. As with most vehicles of this type, the doors, windscreen and top have been removed. (*Simon Thomson*)

WMIK-equipped Wolf XD Defender patrol vehicles generally also feature a swivel mount on the left-hand side of the body to accept a general-purpose machine gun (GPMG). Note the canvas covers fitted over the headlamps to reduce reflections. (*Warehouse Collection*)

A well-stowed Wolf XD Defender equipped with MILAN wire-guided anti-tank missile launcher, WMIK, rear stowage basket and deep-water snorkel. The distinctive alloy wheels are typical of late model WMIK Wolf Defenders. (*Warehouse Collection*)

LAND ROVER AUSTRALIA
Perentie MC2 and MC2HD Land Rover

The Australian Defence Force (ADF) had been using locally produced Series IIA Land Rovers since the early 1960s before moving on to the Series III. In 1981 Jaguar-Rover Australia (JRA) produced a heavy-duty 3-ton 6x6 truck using components of the Defender 110, and based on a development vehicle from SMC Engineering of Bristol. At more or less the same time the ADF announced its intention to procure 25,000 1-ton and 400 2-ton vehicles under 'Project Perentie'. Seven companies submitted tenders, and three, including JRA and Jeep, provided vehicles for trials; at the conclusion of these trials JRA received contracts for both types of vehicle.

Based on a standard Defender 110, the 1-ton variant was designated the MC2 4x4 Perentie, with deliveries taking place between 1987 and 1990. The chassis was produced as a soft-top cargo vehicle, hard-top and soft-top communications vehicles, personnel carrier/station wagon, command vehicle and uprated 1.2-ton surveillance vehicle.

The larger vehicle, known as the MC2HD 6x6 Perentie, was constructed on a purpose-designed heavy-duty galvanised-steel chassis, with a fabricated rear section of rectangular tube to support the rear bogie. Power came from an Isuzu 4BD1 four-cylinder turbocharged diesel engine, driving through the four-speed transmission of the early Range Rover; drive to the rear-most axle was by a separate propeller shaft from the transfer box power take-off. The front and rear axles were wider than standard, and were fitted with lower ratio gears to accommodate the increased weight. Deliveries of the MC2HD started in March 1989, and the vehicle was also offered worldwide, with the option of a 3.5-litre V8 petrol engine. Standard variants included a pick-up truck, personnel carrier, water tanker, air-defence vehicle, gun tractor, four-stretcher ambulance, box body and shelter carrier.

Both versions were also used as the basis of a very effective long-range patrol vehicle for the Australian SAS. The smaller 4x4 version resembled the British SAS desert patrol vehicle (DPV), while the extra length of the 6x6 provided considerably enhanced carrying capacity.

The Perentie MC2HD 6x6 is a heavy-duty Land Rover constructed by Jaguar-Rover Australia (JRA). Although originally intended as a cargo vehicle and tractor, the chassis was also used to produce a well-armed long-range patrol vehicle for the Australian SAS. (*Australian Defence Force*)

To the left is the 1-ton Perentie MC2 4x4 vehicle modified for long range patrol duties, while alongside are two examples of the 2-ton MC2HD 6x6 patrol vehicle. (*Australian Defence Force*)

Men of the Australian SAS patrol in a well-equipped Perentie MC2HD vehicle; as seems to have become the norm with such photographs, their faces have been blurred to prevent recognition. (*Australian Defence Force*)

Mechem BAT

Produced by Denel (Pty) of Silverton, South Africa, and in service with the South African National Defence Force (SANDF), the Mechem BAT is a lightweight Special Forces utility vehicle that has been developed, as far as possible, using civilian automotive components. There is a choice of either a 3-litre V6 petrol engine or a 2-litre four-cylinder turbocharged diesel, both arranged to drive all four wheels through a five-speed gearbox and four-speed transfer case, giving a top speed of 75mph. Live axles are suspended on semi-elliptical leaf springs, with telescopic hydraulic shock absorbers, and a cable-operated differential lock is fitted to the rear axle. The open-topped steel-framed body carries armoured protection to the lower areas, and there is an armoured glass windscreen, which includes a central twin machine-gun mount. The rear of the body is constructed to accept a range of easily mounted palletised modular inserts: these include a six-seat personnel module; three-seat personnel module with 60mm mortar, or 68mm or 107mm rocket launcher. Options include a removable cab roof, towbar, differential lock for the front axle and an electric winch.

A Mk 2 version appeared in 2001, using a Mercedes-Benz five-cylinder diesel engine in conjunction with a four-speed automatic gearbox, pushing the top speed up to 85mph on the road. The live axles of the original were retained, but the semi-elliptical springs were replaced by long-travel coil springs. A frontal armour kit was available as an option.

The Mechem BAT is a lightweight Special Forces utility vehicle developed by Denel (Pty) of Silverton, South Africa; the rear of the body will accept a range of easily mounted palletised modular inserts. The vehicle is in service with the South African National Defence Force (SANDF) in a number of roles. (*Warehouse Collection*)

Mechem BAT Mk 2 in service with the South African National Defence Force and equipped with a 60mm mortar; the vehicle can also be fitted with a variety of other weapons. (*Warehouse Collection*)

MERCEDES-BENZ G-WAGEN IFAV

Although it is invariably credited to Mercedes-Benz, the off-road vehicle that is universally described as the G-Wagen was actually developed by the Austrian company Gelaendewagenfahrzeug Gessellschaft (GFG) in 1979. It was inspired by proposals for a military vehicle developed by Shah Mohammad Reza Pahlavi of Iran in 1976, but was first sold on the civilian market. At the time GFG was part-owned by Mercedes-Benz and part-owned by Steyr-Daimler-Puch. When GFG was closed down in the early 1980s, Steyr-Daimler-Puch continued to produce the vehicle under direct contract to Mercedes-Benz, with the vehicle known as the G Class. At the time of the vehicle's launch, there was a choice of two wheelbase lengths, 94in and 112in, but the original shorter wheelbase was soon discontinued, to be replaced by a longer 135in variant. The G-Wagen's military career started in 1980 when Peugeot entered it in a competition held by the French Army to find a replacement for the ageing Hotchkiss M201 Jeep. The Peugeot was selected against competition from Citroën and Renault, and in 1982 entered service as the licence-built P4.

The engine is a five-cylinder common-rail turbocharged diesel engine of 2.8 litres, driving both axles through a four-speed automatic transmission and two-speed transfer box. The suspension, which consists of live axles on coil springs with axle location by longitudinal and transverse links, lacks sophistication, but both front and rear axles include differential locks and the vehicle has acquired a reputation for extreme off-road capability.

Military users can currently choose from four variants: station wagon; van; open cargo vehicle, which can also be configured as a weapons mount; and a chassis-cab for special bodywork, which includes an ambulance for forward areas. A crew cab is also available, as are armoured variants, and both Peugeot and the US company Stewart & Stevenson, the latter now part of British Aerospace, use the G-Wagen as the basis for a Special Forces vehicle, the latter in service with the US Marine Corps where it is described as the 'interim fast attack vehicle' (IFAV).

Since its launch, more than 60,000 Mercedes G Class military vehicles have been supplied to more than twenty-four nations, including Germany, Ireland and the USA. As well as the Peugeot P4, production of which is now complete, the vehicle is also produced under licence by Hellenic Vehicle Industry in Greece, where it is described as the 290GD.

US Marines with 2nd Combat Engineer Battalion, Battalion Landing Team 3/6, 26th Marine Expeditionary Unit loading a Mercedes-Benz G-Wagen interim fast attack vehicle (IFAV) at Kandahar International Airport, Afghanistan, during Operation Enduring Freedom. In the background are the remains of a Soviet Mi-18 (HIP H) helicopter. (*Captain Charles Grow; US Marines*)

Even the best maintained vehicles will sometimes fail ... here, one Mercedes-Benz G-Wagen interim fast attack vehicle (IFAV) of the US Marines is seen towing a second similar vehicle. The location is Kandahar International Airport. Note the tail-pipe extensions. (*Captain Charles Grow; US Marines*)

A privately owned Mercedes-Benz G-Wagen interim fast attack vehicle (IFAV) designed by Stewart & Stevenson. The photograph clearly shows the infrared driving lights, the distinctive mudguard-mounted air intake designed to allow deep-water wading, and the roll-cage. (*Simon Thomson*)

MINERVA LAND ROVER

At the end of the 1940s the Belgian Army was seeking replacements for its ageing American Jeeps and in 1951 agreed to buy an initial 2,500 Land Rovers from a consortium which had been established between Minerva, a Belgian-based car company that had been founded in Antwerp in 1899, and the British Rover Company. Construction started on 12 September 1951, with knocked-down kits supplied by Rover, consisting of the complete rolling chassis, together with the vital bulkhead. The bodies, which featured rather ungainly flat-fronted mudguards and a fixed panel at the rear in place of the standard Land Rover tailgate, were constructed locally from steel rather than aluminium. Lighting equipment, fuel tanks and upholstery came from Belgian manufacturers, with Minerva claiming that there was 60 per cent local content. At first the vehicles were based on the original 80in chassis and were powered by the F-head four-cylinder Rover engine of 1997cc, driving through a four-speed transmission. From 1954 the 86in chassis was used, with some sources suggesting that, for late production vehicles, the chassis was produced in Belgium.

Alongside the standard utility vehicle, there was a 24V fitted-for-radio (FFR) variant with a screened electrical system, a two-stretcher field ambulance, and a Special Forces vehicle modelled on the SAS Land Rovers. Intended for the parachute/commando role, this latter vehicle was armed with three FN MAG 7.62mm machine guns. Other modifications included strengthened suspension, outboard headlamps alongside an armoured shuttered grille, wing-mounted black-out lights, front-mounted spare wheel, armoured-glass aero-type screens, a rear stowage basket and side-mounted grab handles; the doors were removed. In 1980 thirteen Minerva Special Forces vehicles were converted to carry the MILAN anti-tank missile.

From October 1953 the Minerva was also offered to civilians; total military production was 5,921 vehicles, with a further 4,000 sold on the commercial market. Minerva went into liquidation in 1958.

Constructed in Belgium, the licence-built Minerva is based on the British Land Rover Series I machine. This example has been modified to provide a Special Forces vehicle in the style of the British SAS vehicles. Typically the doors are removed and the spare wheel relocated to the front to increase stowage space. (*Warehouse Collection*)

Heavily modified, this Minerva was designed for the parachute/commando role. Modifications include strengthened suspension, outboard headlamps alongside an armoured shuttered grille, black-out lights, front-mounted spare wheel, armoured-glass aero-screens and rear stowage basket. Weaponry usually includes three FN MAG 7.62mm machine guns. (*Warehouse Collection*)

NORDAC NMC-40 WARRIOR FAV

The Nordac NMC-40 Warrior was an experimental three-seat long-range fast attack vehicle (FAV) with seating for a crew of two or three men. In the three-man version, the driver and co-driver were seated at the front, with a machine gun mounted ahead of the co-driver, while the third crewmember faced to the rear, with a second machine gun. Power came from an 1,800cc four-cylinder air-cooled Volkswagen engine mounted behind the rear axle, driving the rear wheels only through a four-speed transaxle. Disc brakes were fitted, with the driver able to select the brakes on one side or the other in order to execute tight, skid turns in the style of an agricultural tractor. The vehicle was trialled by the US Army in the early 1980s, but there was no series production.

Trialled by the US Army in 1982/83 but never purchased in quantity, the Nordac Warrior NMC-40 fast attack vehicle (FAV) had seating for a crew of either two or three men; in the three-man version, the third crew member was placed between the backs of the front seats and the engine, firing his machine gun to the rear across the engine compartment. (*Warehouse Collection*)

PINZGAUER 710

In 1965 the Austrian company Steyr-Daimler-Puch showed a prototype of its Model 710 Pinzgauer 4x4; this was followed by the 6x6 Model 712. By 1971 both were in production. Both models were initially offered with either a fully enclosed steel body or a military-type cargo body with a folding windscreen; the cargo area had removable top bows, with bench seats fitted down both sides of the body for either eight men (Model 710) or twelve men (Model 712). During trials of the vehicle in Britain in 1993, Automotive Technik of Aldershot, Surrey was established to provide trials support and to handle deliveries. When commercial production in Austria ceased in June 2000, Automotive Technik acquired the rights to the design and the production line was shifted to new premises in Guildford. Ownership of the company subsequently passed to Stewart & Stevenson, and then to BAE Land Systems.

The Pinzgauer is constructed around a torsion-resistant tubular backbone chassis on which are mounted independent swing axles with coil springs; there are rocking beam leaf springs at the rear of 6x6 variants. Early examples used a Steyr four-cylinder petrol engine but from 1983 the petrol-engined versions were joined by a turbocharged 2.4-litre six-cylinder diesel engine; subsequently, a Volkswagen turbocharged 2.4-litre five-cylinder diesel engine has also been specified. The engine is coupled to a five-speed ZF gearbox and two-speed transfer box, with lockable differentials front and rear. A four-speed ZF automatic gearbox is also available. An 'improved medium mobility' variant of both the 4x4 and 6x6 configurations was released in 2002, and there is also a so-called 'extreme mobility' variant, with electronic traction control.

Alongside the basic cargo/personnel vehicles, the Pinzgauer has been produced for the command, flat top and shelter carrier, weapons carrier, radio, workshop, fire-fighter and ambulance roles. The Special Forces variant, described as an air-portable weapons platform and designated 'WP', was developed by Ricardo Special Vehicles and is available in both 4x4 and 6x6 configuration. A combined roll-cage and weapons mount is fitted, in the style of the Land Rover WMIK, allowing a heavy machine gun to be mounted over the rear compartment; a 7.62mm general-purpose machine gun is carried on a side pintle mount ahead of the co-driver.

Developed by Ricardo Special Vehicles for the Special Forces role, the Pinzgauer WP (weapons platform) is an air-portable vehicle available in both 4×4 and 6×6 configuration. The WMIK-style combined roll-cage and weapons mount allows a heavy machine gun to be mounted over the rear compartment. (*Warehouse Collection*)

Like the WMIK-equipped Land Rover Wolf XD, the Pinzgauer WP also mounts a 7.62mm general-purpose machine gun on a side pintle mount ahead of the co-driver. (*Warehouse Collection*)

The Pinzgauer WP was produced in both 4x4 and 6x6 format; thirteen of the latter were purchased by the New Zealand Defence Force in the manufacturer's XM extreme mobility configuration for use by Special Forces. (*Warehouse Collection*)

SMAI LWV

Constructed by the Argenteuil-based French company Soudure et Mécanique Appliquée Industrielles (SMAI) in the late 1980s, the LWV is a low-cost, lightweight (1,760lb), Jeep-like vehicle designed for use by airborne and Special Forces, and created with ease of maintenance and production in mind. Customers are able to choose from a range of diesel or petrol engines, including four-, five-, six- and eight-cylinder units; there is a similar choice of transmission, driving all four wheels through either a manual or automatic gearbox. Rigid axles are carried on coil springs with hydraulic telescopic shock absorbers, and axle location is by means of Panhard rods. By 1992 it was said that development was complete and the vehicle was ready for production. Typical roles include troop carrier, ambulance, anti-tank missile platform, combat support vehicle, reconnaissance, command and rapid intervention.

Scale model of the French SMAI LWV, which was developed in versions for troop carrying, ambulance, anti-tank missile platform, combat support, reconnaissance, command and rapid intervention roles. (*Warehouse Collection*)

ST KINETICS FLYER R-12D

Originally developed by the US HSMV Corporation but subsequently taken over by Singapore Technologies Kinetics (STK), the R-12D is a buggy-style light strike vehicle (LSV) constructed around a typical tubular space-frame chassis, with an integral roll-cage and weapons mount designed to accept a heavy machine gun, 40mm grenade launcher or missile launcher. Power comes from a rear-mounted 2-litre turbocharged diesel engine driving all four wheels through a three-speed semi-automatic transaxle. Suspension is independent at all four wheels.

The R-12D is in service with the Singapore armed forces, but has been replaced in production by the more powerful Spider LSV.

The R-12D was developed by the US HSMV Corporation, and was manufactured by Singapore Technologies Kinetics (STK) before being replaced by the similar Spider light strike vehicle (LSV). The R-12D shown here was exhibited at a Far East defence trade show equipped with the STK 120mm 'super rapid advanced mortar system'. (*Simon Thomson*)

The R-12D is a buggy-style light strike vehicle constructed around a tubular space-frame chassis and equipped with a rear-mounted diesel engine. The vehicle is in service with the Singapore armed forces. (*Warehouse Collection*)

SUPACAT ATMP

Originally developed by Devon-based Supacat Ltd in 1982 for agricultural, utility and other civilian applications, it soon became clear that the versatility of the Supacat made it an ideal military vehicle and the first examples of what was described as the Supacat 6x6 all-terrain mobile platform (ATMP) entered military service in 1984. A decade later the company reached an agreement with Alvis to produce and market the vehicle, although production is now back in the hands of the original designers, with manufacturing assistance from Devonport Management Ltd (DML), part of the Babcock International Group.

The six-wheeled Supacat is a multi-purpose, multi-role vehicle which is eminently suitable for the Special Forces role; as well as being adapted as a weapons or missile platform, light gun tractor, light recovery vehicle or cargo/personnel carrier, it can also be fitted with a crane, hydraulic digger or winch, and can be readily converted to amphibious operation using an outboard motor. The combination of wide, low-pressure tyres, all-wheel drive and a driver-selected choice of Ackermann or skid steering provides excellent mobility and the vehicle is also fully amphibious. Constructed around an aluminium-clad steel space-frame forming an enclosed hull, driver's compartment and load-carrying area, the Supacat is powered by a Volkswagen 1,900cc turbocharged diesel engine, mounted amidships, driving all six wheels through a four-speed automatic gearbox; drive to the forward and rear axles is made via a duplex roller chain-drive system.

In its standard form the Supacat is a two-seat open vehicle with minimal bodywork and without a windscreen or cab. A range of optional enclosures is also available, including a full hardcab, soft-top or removable canopy; the Special Forces variant is fitted with a combined roll-cage and weapons mount. The small size of the vehicle means that it can be carried in standard fixed-wing transport aircraft, with the vehicles stacked one on another; for example, eight vehicles can be carried in a C-130 transport aircraft. It can also be under-slung from a number of standard military helicopters; up to four vehicles can be carried under a Chinook using a suitable platform, two vehicles by a Blackhawk, and single loads can be under-slung from Puma and Sea King helicopters.

On the road the Supacat has a top speed of 40mph, and can climb a gradient of 45 degrees and traverse a side slope of 40 degrees. The standard payload is 3,520lb but this can be increased to 8,800lb under certain conditions by the use of a special two-wheeled self-loading trailer described as the 'fork lift pallet trailer' or FLPT; a standard 1-ton two-wheeled box trailer is also available.

The Supacat is currently in service in Britain with 16 Air Assault Brigade, the Royal Marines, 5 Airborne and 24 Air Mobile. It saw service in the Gulf during Operation Desert Storm and was air-dropped into the Balkans to assist with UN humanitarian relief operations. Overseas operators include units of the Canadian and Mexican armies, and a number have been ordered by an un-named Far Eastern country for deployment as a rapid clearance vehicle in air-drop zones.

Lacking any kind of weather protection, and with a minimal superstructure, the Supacat all-terrain mobile platform (ATMP) is a six-wheeled multi-purpose, multi-role vehicle which is used by Britain's 16 Air Assault Brigade, the Royal Marines, 5 Airborne and 24 Air Mobile. It is powered by a Volkswagen turbocharged diesel engine driving all six wheels through a four-speed automatic gearbox, with final drive by chain to the front and rear axles. (*Warehouse Collection*)

Designed for a standard payload of 3,520lb, which can be increased to 8,800lb if necessary, the Supacat ATMP is a versatile and highly accomplished machine capable of climbing a gradient of 45 degrees and with a top speed of 40mph. (*Warehouse Collection*)

SUPACAT HMT 400 JACKAL

Officially described as the MWMIK ('mobility weapons mount installation kit'), the Supacat 'high mobility transport' (HMT 400) Jackal was developed to provide a well-protected, fast and agile patrol vehicle for the British Army. Designed by Supacat and manufactured in volume by Babcock International under licence from Lockheed Martin, it was the first of what should probably be considered as the third generation of Special Forces vehicles to be ordered in quantity, with the first examples entering service in Helmand Province in 2008. Initial orders covered 172 vehicles, but in 2009 it was announced by the Ministry of Defence that another 110 improved Jackal 2 vehicles were being ordered, with a further 140 blast-protected Jackal 2A variants coming a year later. It was said that eventually this would see the total number of Jackals in service reach 500. The vehicle has been adopted by the SAS as a replacement for the ageing Land Rover Defender 110-based special operations vehicles.

The Jackal is designed to carry a crew of four or five men. Its larger size enables a payload of 4,620lb to be carried, but its top speeds of 75mph on the road and 50mph across country remain comparable to those of the Land Rover and the vehicle is still sufficiently compact to be able to fit inside a CH-47 Chinook helicopter. An optional mine blast and ballistic protection kit is available, designed by Jankel Engineering, and the vehicle can also be fitted with a range of mission-specific stowage hampers, weapons, and communications and ISTAR equipment (intelligence, surveillance, target acquisition and reconnaissance) to suit a wide range of operational roles. The Jackal has already been deployed in the harsh terrain of Afghanistan, where it has acquitted itself well.

Power comes from a front-mounted Cummins six-cylinder turbocharged diesel engine; the early versions were fitted with a 5.9-litre engine, but the Jackal 2 has a more powerful 6.7-litre unit producing 180bhp. The engine is coupled to all four wheels via a five-speed automatic gearbox and two-speed transfer box, and both axles are fitted with limited slip differentials. There is independent air-operated spring suspension at all four wheels using wishbones and twin shock absorbers; the air springs provide an air-adjustable ride-height facility that offers exceptional ground clearance; a very real benefit of this sophisticated suspension is that it also provides a very stable firing platform.

The main armament is a 12.7mm heavy machine gun, with the 7.62mm general-purpose machine gun (GPMG) as a secondary weapon. Options include an electric winch, run-flat tyres and fully locking differentials.

The SAS is also currently deploying the six-wheeled Supacat I IMT 600 – or Coyote – which is derived from the Jackal and shares automotive components; the Coyote offers a carrying capacity of 8,580lb for a gross vehicle weight of 23,100lb. There is also a third version of the vehicle, described as the HMT Extenda, that has been developed specifically for use by Special Forces. This is designed to facilitate easy conversion from 4x4 to 6x6 configuration simply by fitting or removing a self-contained modular third axle unit. The same range of options is available, including the ballistic-protection kit.

Designed by Supacat, but with volume production undertaken by Babcock International, the HMT 400 Jackal is a 4×4 Special Forces vehicle designed for a crew of four or five. The combination of a Cummins turbocharged diesel engine and air-adjustable suspension provides excellent top speed and good cross-country performance. (*Simon Thomson*)

The MWMIK-equipped Jackal is being used by the SAS to replace the ageing Land Rover Defender 110-based special operations vehicles. This vehicle was photographed at Camp Bastion, Afghanistan. (*Corporal Ian Houlding; MoD Crown copyright*)

The HMT 600 Coyote is effectively a 6x6 version of the smaller Jackal with enhanced load-carrying abilities. These examples were photographed during an eight-day training course at Driffield Training Area, Leconfield, East Yorkshire. (*Andrew Linnett; MoD Crown copyright*)

SZÖCSKE LSV

Development of the Hungarian GÉPFET Szöcske (the name simply means 'cricket') light strike vehicle started in 1992 on behalf of the Hungarian Ministry of Defence. The machine is designed around a tubular space-frame, which doubles as a roll-cage and weapons mount, and is powered by a rear-mounted 1.9-litre petrol engine, driving all four wheels through a five-speed manual gearbox that includes a low-ratio climbing gear. The suspension is independent, using trailing arms and coil springs.

Designed to be operated by a crew of four, the vehicle can be used to mount a recoilless rifle or anti-tank missile launcher.

Work started on the Szöcske light strike vehicle in 1992 at the request of the Hungarian Ministry of Defence, although it is not clear whether any examples were actually purchased. The vehicle was powered by an air-cooled petrol engine, with both two- and four-wheel drive versions available; like the WMIK-equipped Land Rovers and most of the buggy-style vehicles, the roll-cage doubled as a weapons mount. (*Warehouse Collection*)

TELEDYNE LIGHT FORCES VEHICLE

Dating from the mid-1980s, the American Teledyne Continental 0.75-ton light forces vehicle (LFV) was designed to provide high speed combined with excellent cross-country performance, and the ability to act as a weapons platform. Power was provided by a turbocharged diesel engine, mounted at the rear, and driving all four wheels through a three-speed automatic gearbox; both axles incorporated limited slip differentials. Braking was by power-assisted discs all-round and there was independent suspension at all four corners using air springs. The vehicle was low and light in weight, with a 70mph top speed and a 0–30 acceleration time of 4.5 seconds. The roll-cage incorporated a ring mount for a heavy machine gun, and the vehicle could also be used to mount, for example, a wire-guided anti-tank missile system.

Just two examples were constructed for evaluation by the US Army, but there was no series production.

The Teledyne light forces vehicle (LFV) dates from the mid-1980s and was equipped with a turbocharged diesel engine driving all four wheels through a three-speed automatic gearbox with limited slip differentials at both ends. It was designed to provide excellent high-speed cross-country performance and the ability to act as a weapons platform, with the roll-cage doubling as a machine gun mount. There was no series production. (*Warehouse Collection*)

WESSEX SAKER LSV

Designed by Wessex plc of Reading and constructed for the British SAS by Devonport Management Ltd (DML) in the late 1980s, the Saker – sometimes called Eagle – is a dune buggy-style fast attack or light strike vehicle constructed around a welded tubular space-frame, with minimal bodywork of aluminium, glass fibre or Kevlar composite armour. The standard power unit is a Volkswagen 1,600cc air-cooled four-cylinder petrol engine driving both the front and rear axles through a four-speed transmission, but the company also offered the option of a Perkins Prima 80T turbocharged diesel engine. All four wheels are independently suspended using a combination of torsion bars and coil-over hydraulic shock absorbers.

Alternative versions of the Saker have also been produced, designed for a crew of either two or four men, with a maximum load-carrying capacity of 1,540lb, and the vehicle can be easily adapted to suit a number of roles, including reconnaissance, long-range incursions, airfield security, etc. Total payload is 1,540lb, and typical vehicle-mounted weapons include a 7.62mm general-purpose machine gun (GPMG), 0.50in Browning heavy machine gun, 30mm ASP cannon or Hellfire anti-tank missile. The area over the rear-mounted engine is designed to be used for stowage, and additional stowage panniers can be attached to the sides of the body. Optional equipment includes run-flat tyres, bush guards, long-range fuel tanks and communications and navigation equipment.

Licencing deals for the production of the Saker were agreed in the USA and Singapore.

The Wessex Saker is a buggy-style fast attack or light strike vehicle of space-frame construction, with minimal bodywork of aluminium, glass fibre or Kevlar composite armour. Alternative two- or four-men versions were produced, with a maximum load-carrying capacity of 1,540lb. The vehicle can be easily adapted to suit a number of roles, including reconnaissance and long-range incursions. (*Warehouse Collection*)

The standard version of the Wessex Saker was powered by a Volkswagen air-cooled four-cylinder petrol engine driving all four wheels via a tough four-speed transmission, and there was independent suspension all round. The Saker can be used to carry a 7.62mm general-purpose machine gun (GPMG), 0.50in Browning heavy machine gun, 30mm ASP cannon or Hellfire anti-tank missile. (*Warehouse Collection*)

OTHER VEHICLES

There are other vehicles that have been produced for Special Forces which would have been described by the authoritative Janes Military Vehicles and Logistics directory as being 'available', meaning that at the time of publication the vehicle had not managed to find a ready market. Examples of these include the French Giat VRA buggy from 1992 and the ACMAT VLRA long-range patrol vehicle. The Romanian Turbomeccanica Company produced a special-purpose dune buggy-style machine dubbed Hamster. Other British vehicles include the Austin Champ (FV1801) which was fitted out with weapons, armoured screens and a recoilless rifle for use by the SAS in the mid-1950s, although it is doubtful that any saw service; there was also the Land Rover-based BIRST MWP fast strike vehicle and the Esarco 6x6, the latter being designed for the British Army's all-terrain mobile platform (ATMP) role that resulted in the acquisition of the Supacat. The Royal Marines are users of the Hagglünds BvS10 Viking, a Swedish-built light armoured tracked vehicle that started to enter service in July 2003. In Israel the AIL Desert Raider was an interesting 6x6 reconnaissance and fast-attack vehicle with a walking-beam rear axle arrangement. Resembling a rather angular HMMWV, the Italian ARIS VAT from 1998 was built on the chassis of the Swiss Bucher DURO truck. And finally, from the USA, there was the HSMV R-12 high-speed mobility vehicle and the TPC Logistics RAMP-V rapid multi-purpose vehicle.

It is also as well to remember that other types of vehicle may well have been adapted to this role. For example, the larger Special Forces units such as the US Marine Corps, the French Foreign Legion and the British Air Assault Brigades have certainly deployed more conventional military vehicles during clandestine missions, including tanks, infantry fighting vehicles, armoured cars and quad bikes or motorcycles . . . the US Marines' Canadian-built General Dynamics light armoured vehicle (LAV), for example, is a well-armoured eight-wheeled reconnaissance vehicle that, in some circumstances, can be considered to be 'special operations capable'. The US Marines also had a fast attack vehicle based on the ageing M151A2 and this remained in service long past its 'sell-by date' simply because it could fit into Marine Corps helicopters, while the wider HMMWV could not.

Even the ubiquitous Toyota pick-up, so beloved of rebel fighters across the developing world, is the perfect chameleon-like vehicle, and is perfectly capable of mounting formidable firepower, ranging from heavy machine guns to anti-aircraft weapons.

Dating from the mid-1950s, this stripped-down, lightly armoured and well-armed Austin Champ (FV1801) may well have been trialled for the SAS before the Land Rover Series I was adopted for the role. The tubes in the rear of the body are components of the US Army's 3.5in recoilless anti-tank rocket launcher; the machine gun is the venerable water-cooled Vickers medium. (*Tank Museum*)

Opposite, top: The US Marines' LAV is an 8x8 light armoured amphibious reconnaissance vehicle constructed by General Dynamics Land Systems, Canada and based on the Swiss MOWAG Piranha. The photograph shows Marines with Company D, Light Armored Reconnaissance, Battalion Landing Team 3/1, 11th Marine Expeditionary Unit, a 'special operations capable' unit. (*Zinnmann; US Marine Corps*)

Opposite, bottom: Based on the Hagglünds Bv206, the BvS10 Viking is a light armoured vehicle designed for the Royal Marines. Its first operational deployment was with Royal Marine Commandos in southern Afghanistan. (*Warehouse Collection*)

In the same way that not every volunteer who submits himself for selection as a Special Forces soldier makes it through the selection process, so there are also vehicles that are proposed or trialled for the role but which are never purchased in quantity. One such was the British Esarco, a 6x6 vehicle manufactured by Laird of Anglesey; among others, it was proposed for consideration as the British Army's all-terrain mobile platform (ATMP), but was eliminated from the selection process. (*Simon Thomson*)

In truth, there is no more a typical Special Forces vehicle than there is a typical Special Forces mission ... it's simply a matter of what gets the job done. Here, an Abrams M1A1 main battle tank of the US Marine Corps' 1st Tank Battalion is photographed in Helmand Province, Afghanistan ... and don't let anyone tell you that the US Marines are not Special Forces. (*Sergeant Jesse Johnson; US Marine Corps*)